Your Guided Tour to America's
Employment Crisis

WHERE DID THE JOBS GO-AND HOW DO WE GET THEM BACK?

Your Guided Tour to America's
Employment Crisis

SCOTT BITTLE AND JEAN JOHNSON

wm

WILLIAM MORROW
An Imprint of HarperCollins*Publishers*

HarperCollins books may be purchased for educational, business, or
sales promotional use. For information please write: Special Markets
Department, HarperCollins Publishers, 10 East 53rd Street, New York,
NY 10022.

FIRST EDITION

Library of Congress Cataloging-in-Publication Data is available upon
request.

ISBN 978-0-06-171566-2

12 13 14 15 16 OV/RRD 10 9 8 7 6 5 4 3 2 1

"Let's work the problem, people. Let's not make things worse by guessing."

NASA flight director Gene Krantz
(played by Ed Harris) in *Apollo 13*

CONTENTS

Contents

CHAPTERS 12–15: STUFF HAPPENS
THE BIG TRENDS THAT COULD AFFECT JOBS

PREFACE

If politics were a Scrabble game, the jobs issue would be the letter *E*.* It pops up more than just about any other issue the country faces. Elections are won and lost over it. Op-ed writers opine about it. Pollsters interrupt family dinners to ask people how they feel about it. From the White House to city halls, politicians have been speechifying about jobs and unemployment almost incessantly for the last several years. Even when they are talking about other problems, such as the federal budget deficit or energy, they still bring it back to employment: their ideas will create jobs, while whatever their opponent suggests is a "job killer." Nearly all of them claim they've got the right idea for reducing unemployment and getting the economy back to abundant jobs and rising salaries. *Just do things my way,* they all seem to say, *and everything will be fine.*

THE NUMBERS ARE ABYSMAL

There's a good reason why politicians are talking jobs. Since

---

* The English version of Scrabble has 100 letter tiles, and 12 of them are *E*'s, more than any other letter. *A* and *I* follow at 9 apiece: www .blurtit.com/q463995.html.

the global financial crisis and the Great Recession began in late 2007, the jobs picture has been so abysmal that it's demoralized and discouraged nearly everyone.[1] Millions of Americans have seen their financial lives unravel. Some will probably never get back to where they were.

- The U.S. economy lost more than 8 million jobs in the recession.[2] The Congressional Budget Office (CBO) estimates that we won't get back to anything near full employment until 2016.[3] Some economists say we face an even longer-term jobs drought.[4]
- In any given month since the recession hit, some 13 to 15 million Americans have been out of work.[5]
- And in any given month, more than 8 million people who wanted full-time work were in part-time jobs instead.[6]
- More than 1 American family in every 10 has someone unemployed.[7]
- Even people with jobs have been taking a hit. Since the recession, 28 percent of Americans say they have had their hours cut back.[8] About a quarter have had their pay reduced. Another 12 percent report being furloughed or asked to take leave—that's roughly one American worker out of ten.[9]

If we were talking about a disease, this would be a pandemic.

A PROBLEM THAT'S HERE TO STAY

What's worse, nearly all the "smart people" (including many who actually are smart) say the jobs situation isn't going to turn around fast. Experts expect unemployment to remain high for years even as other aspects of the economy may be improving.

Given the head-spinning changes in the world economy over the last ten or fifteen years, we may well have come to the moment when this country has to start doing some things very differently if we're going to have enough jobs for roughly 154 million American workers.[10] Since the vast majority of us rely on a job (or jobs) of some kind to put food on the table, this is hardly some abstract, far-off problem for people in posh hotel conference rooms in Washington, D.C. This is something all of us need to understand.

But having *enough* jobs to go around isn't enough—at least we don't think so. There's also the question of whether there are enough jobs here in the United States that pay enough for workers to maintain a decent standard of living and still have a little money over at the end of the month. A lifetime spent in jobs that pay peanuts, skimp on benefits, and vanish into thin air every time the economy hits a bump isn't exactly what most Americans have in mind.

CAN WE GET BACK TO WHERE WE BELONG?

Sensing, rightly, that voters are deeply unnerved by what's been happening to American jobs, politicians are scrambling. They have been tossing out ideas for "stimulating" the economy, creating new jobs, saving old ones, and getting us back to where we were before the great economic meltdown.

The jobs issue is poised to be a major theme in the 2012 presidential election, and other campaigns too. It's hard to imagine that anyone wouldn't like to vote for candidates for president, Congress, governor, and mayor who will make this issue a genuine priority and will also make smart, compassionate, forward-looking decisions about it—decisions that set the United States on a good path for jobs for many years to come. But given what goes on in campaigns these days—the sloganeering and the posturing, the distortion and the half-truths, the snake oil and the hyperbole—deciphering what politicians are really saying and what they'll actually do once they're in office can seem like a full-time job itself.

And that is why we wrote this book. The country faces a grave problem—one that can undermine the lives of millions of Americans—and the political discussion about it is, frankly, horrible: opportunistic, confusing, and a mishmash of bombast and denial. Our book is designed to help voters sift through the political rhetoric and begin to sort out the clatter of proposals politicians are offering. It's intended to provide some context and clarification on an issue that is, for most of us, confusing and complicated. We also want to make sure readers get a chance to think about some ideas and trends that *aren't* being raised by politicians. These are ideas that aren't easily packaged into sound bites, but some of them could be crucial to turning the U.S. jobs picture around.

Where Did the Jobs Go? is a citizens' guide to the jobs issue, written specifically for readers who aren't economists, financiers, business school professors, or policy wonks working for think tanks. In the following pages, we'll explain some of the best thinking about what's been happening to American jobs and lay out the major options for addressing the problem as clearly as we can. We'll cover proposals from the political left, right, and center and try to help you understand risks, costs, and trade-offs associated with each of them. We have tried to ditch the economists' lingo and keep the data dump to a minimum.

WHY US?

We've written similar books before. One of our books—
*Where Does the Money Go? Your Guided Tour to the Federal
Budget Crisis*—looks at the government's hemorrhaging
debt and the financial dangers facing Social Security
and Medicare. Another, *Who Turned Out the Lights? Your
Guided Tour to the Energy Crisis*, spells out lots of choices
for changing the way the country gets and uses both traditional and alternative energy.

We are both senior fellows at Public Agenda, a nonprofit,
nonpartisan research organization that covers policy issues
and conducts public opinion research.* In our work there, we
have spent years translating expert information into terms
and concepts that people who aren't experts can understand.
This series of books is an outgrowth of that work.

Now that we're tackling the jobs issue—and before getting to the nuts and bolts—we'd like to explain a little of our
thinking behind this book:

**First, we're going to focus expressly on jobs—how to cre-
ate them and keep more good ones in the United States.**
This is not a book about how to generate economic growth
in general or how the U.S. economy can compete better
globally. It's not about how to shrink the gap between the
rich and poor or how to reduce poverty. Those issues are
important. They are all related to the jobs issue, and we'll
certainly touch on them as we go along. But our theme is
jobs, jobs, jobs. It's the public's number one measure of
whether the economy is working, so our intent is to sort

---

* You can find out more about Public Agenda at www.publicagenda
.org.

through what the country might have to do (or avoid doing) to get better results on that score.

Second, we're going to take a look at the long term—not just the problems of the Great Recession.

For the last couple of years, the United States has been digging out of one of the worst recessions since the Civil War. (There have actually been more than thirty of them since that time,[11] although back then they were known as "financial panics" rather than "recessions." But these days even economic downturns seem to have spin doctors.) Since the economy went into a tailspin in the second half of 2008, there has been a continuing debate about how to perk it back up and stem job losses in the meantime. Economists and politicians have spent so much time talking about stimulating growth and job creation that they sometimes seem like bartenders pushing coffee at closing time, hoping the economy will wake up and get moving again. As this book goes to press, the United States is technically *not* in recession and the economy is growing, but the jobs picture is still dismal for millions and millions of Americans. Moreover (and we'll come back to this in more detail later), even *before* the Great Recession hit, the jobs situation left a lot to be desired. That's why we'll be looking at some longer-term issues and solutions—issues that will affect the kinds of jobs we have here over the next couple of decades.

Third, we're taking the toolbox approach.

Maybe you have a friend or relative who can spend hours happily wandering around Sears or Home Depot. These people (and maybe you're one of them) seem to know what all the various tools and devices do and apparently get a kick out of examining and comparing them. But if you're not a Mr. or Ms. Fix-It, a visit to the hardware department is a mind-numbing experience. Luckily, there is an option. You can buy one of those little red metal boxes containing your basic hammer, pliers, and flathead and Phillips screwdrivers, along with a selection of the nails and screws most people need most often. That's essentially what this book does. We've sorted through countless books, reports, analyses, and articles about jobs and the economy and tried to select and organize the most crucial facts and ideas and put them together in one place. This book doesn't contain everything you could possibly need—even with a toolbox, you may have to run out to the hardware store to get something specific—but we hope it's a good starting point.

Fourth, our mission is to explain the basics.

Neither of us is an economist, and although we've spent our careers covering many different public policy issues, we can't claim that we started this book with any particular expertise on jobs and joblessness. However, we've done the homework, and as strange as it sounds, not being an expert is an advantage when it comes to translating a complex, convoluted, and contentious issue for typical citizens. When we set out to write this book, we came to the jobs issue with the same questions, doubts, suspicions, and moments of complete befuddlement that other Americans have. Having put in the time to sift through the facts and sort out the arguments, we'd like to think we can help you do it too.

Fifth, we're presenting choices, not recommending solutions.

Haranguing the rest of us about how to solve the country's economic problems does seem to be one area where plenty of people are finding work. There are "experts" in newspapers, on cable news and business shows, and all over the Internet who will tell you exactly what should be done about the U.S. economy and jobs. Most of them seem smart, a lot of them are very well dressed, and they all are positive they've got the answers. Still, it's worth remembering that former Federal Reserve chairman Alan Greenspan—who was touted for years as someone who could feel changes in the business cycle the way animals sense earthquakes coming—missed the whole Wall Street bubble, and he wasn't the only economic biggie to miss that one. So maybe it's time for more of us to do more thinking for ourselves. Our aim is to help you do just that. Instead of recommending solutions, we'll give you some background information, and then we'll describe a broad range of choices and summarize the pros and cons. We list our sources so you can dig into this more if you want to. The rest is up to you.

Sixth, we're writing a guide for citizens, not offering advice for investors, entrepreneurs, or job hunters.
It's probably obvious from what we've said so far, but the focus of *Where Did the Jobs Go?* is on the public decisions that national, state, and local governments make and on the broad social, political, and economic developments that surround them. We'll home in on the things that government might do and things it ought to avoid doing to improve the jobs picture for the future. Based on our experience, many Americans dramatically overestimate how much influence the president, Congress, or their state's governor really has on jobs and unemployment. The United States has a private economy, and most of what happens reflects countless individual decisions made across the country by consumers, investors, business owners, bankers, inventors, and others. Nevertheless, government policies on taxes, investment, financial markets, immigration, trade, education, transportation, energy, and other arenas affect how many jobs we have in the United States and whether they're the kinds of

jobs that nurture and support a strong middle class. That's what we'll be writing about.

Seventh, small things change, but the big questions remain.

There are researchers in government and the private sector who measure just about everything in our economy that can be measured—economic growth, productivity, imports, exports, unemployment, how many new jobs the economy is creating, and more. These numbers change all the time. That's why all those charts tracking economic data have so many little squiggles in them. Maybe there will be better news on the jobs front after this book goes to press. We sincerely hope so. But upticks, however welcome, shouldn't distract us from facing up to the big, big, big challenges the United States faces on jobs.

To keep up with population growth, we need more jobs. To make up for the millions of jobs lost in the recession, we need more jobs. To maintain our standard of living, we need jobs that pay good wages. To be a nation that truly offers a fair chance for everyone, we need jobs for people who were falling through the cracks even in good economic times. To ensure that our children inherit a country that still offers the American dream, we need to make decisions that will provide them with good jobs and opportunities too.

Eighth, there is no foolproof formula for getting the good jobs back fast.

We'll say it right up front: this is the toughest issue we've ever tackled. Other issues come down to arithmetic. To balance the budget, you can raise taxes, cut spending, or do some of both. On energy, moving away from fossil fuels and developing alternatives is expensive and inconvenient, but most of the engineers and scientists say it's doable. There are realistic options out there. With jobs, sad to say, there is no surefire equation. There are certainly things the country could do to better its chances of creating more jobs and preserving the jobs it has, and there are certainly things we need to avoid, but in some respects it's a little like love. There is no recipe—when it goes right, it seems magical, and when it doesn't, you just have to pick yourself up, try to figure out what went wrong, and start all over again.

One goal we have in writing this book is to insert a little humility into this whole discussion about jobs. Experts and politicians may be very articulate and persuasive, but that doesn't mean that their recommendations will produce a surefire success. They don't have any iron-clad answers.

So this book is about the choices Americans need to make to address our jobs challenge, and your thoughts and ideas are as important to this national discussion as those of any politician or talking head.

HOW BIG AN ISSUE IS THIS?

Some people may ask how much of a problem this jobs thing really is. After all, most people who lose their jobs find a new one after a while. Maybe it's not exactly what they were looking for, and maybe it doesn't pay as much as their old job or offer as much promise; still, most people who look hard enough for work do eventually find a job of some sort. There are even a couple of interesting books from people who went from working in posh corner offices to working behind a counter and found some good in their changed circumstances.* People adjust. If needed, they

---

* See, for example, *How Starbucks Saved My Life: A Son of Privilege Learns to Live Like Everyone Else* by Michael Gates Gill. Of course, he did feel the need to supplement his Starbucks earnings by writing a book, which we assume he didn't do for free. And where is he now? We'll let you check that out for yourself. It's what the Internet is for.

downsize their dreams and lifestyles. The United States has a huge economy and incredible resources. Even if the jobs situation isn't what it once was, we're still doing better than the vast majority of humankind.

But is that enough? When adults who aren't especially good with children are forced to talk to one, they often resort to asking: "So what would you like to be when you grow up?" The question itself implies that nearly all adults work and that the kind of work you do is part of your identity. But it also implies that we live in a country where children have choices about what they do in life. It suggests that ours is a country where people can shape their own future and find work that's meaningful and productive—work that's just right for them.

It's at the very core of how Americans see themselves: If you work hard, you'll thrive. If you have aspirations, you can pursue them. That's what's in danger now. Figuring out how to reinvigorate that quintessentially American social contract is what this book is about.

|||||||||||||||||||||||||||||||||||||||||||||||||||||||||||||||||||||||||||||||||||

CHAPTER 1

IT'S NOT OVER UNTIL IT'S OVER

Six Reasons to Worry About Jobs Even If You Have One

> So no one told you life was gonna be this way.
> Your job's a joke, you're broke, your love life's DOA.
> —*The Rembrandts, "I'll Be There for You"*

We can't help you with your love life, but we have got some thoughts on the jobs and money part of it. The way most Americans see it, the economy just isn't providing enough good jobs at good wages for people who need and want them.

Most of us hoped that by now the United States would be closing the door on all the economic fallout of the Great Recession. Some trends have improved, but we still face an uncertain, troubled road ahead on jobs. Will we have enough of them? And will they be the kinds of jobs that give people a secure and decent future?

"WHAT ARE THESE? LIKE, FAMOUS CHICKENS?"

We opened the chapter with a couple of lines from the theme song from *Friends*,* and the show isn't a bad place to start. The actors who played the six "friends" certainly did well enough. At the time, they earned some of the highest salaries ever paid to performers in a TV comedy series.[1] But the work lives of the twentysomething characters they played—Chandler, Ross, Joey, Monica, Rachel, and Phoebe—were a mixed bag. In one episode, the six go out to dinner to celebrate Monica's recent promotion. Chandler, Ross, and Monica have good jobs in promising careers. Meanwhile, Rachel, Joey, and Phoebe are bouncing from job to job, barely able to make ends meet. At a posh restaurant, the more prosperous friends enjoy a good meal and generally make merry. Meanwhile, their less securely employed pals nervously scan the menu, where even the chicken dishes are pricey.[2] Outrageously pricey. At one point Joey asks, "What are these? Like, famous chickens?"[3]

This may be one of the few concessions to economic reality during the entire run of the series. After all, even the intermittently employed Joey, Phoebe, and Rachel lived in improbably huge Manhattan apartments. Yet this dinner among friends mirrors the broader experiences of Americans overall. Those of us with good jobs and prospects are enjoying dining out again. Those of us without good jobs or with no job at all are worrying and scrambling to pay the bills.

---

* Our apologies to readers who now have the *Friends* theme stuck in their heads. It came up as number 25 when AOL Radio compiled a list of the worst songs of all time. "Macarena" and that strange "I'm too sexy for my shirt, too sexy for my shirt" song also made the cut: www.aolradioblog.com/2010/09/11/100-worst-songs-ever-part-four-of-five.

LIVE LONG AND PROSPER?

The economic nosedive of 2008 and 2009 jangled everyone's nerves—even Americans who managed to survive with jobs, homes, and savings relatively intact. Businesspeople and entrepreneurs were rattled as well—even very successful ones have been slow to expand their businesses and hire new workers. What many of us don't realize, however, is that according to the numbers, the United States has actually been stumbling for a good decade or more. Here's some of what we're up against:

The country lost millions of jobs in the Great Recession, and it will take years to get them back.

Between the end of 2007 and the beginning of 2010, the U.S. economy lost about 8.4 million jobs.[4] Companies laid workers off, and many folded, so their entire workforces lost their jobs. State and local governments have been slashing payrolls as well.

It could be years before unemployment rates return to prerecession levels—if they ever do.

Before the global economic crisis turned into a tailspin, unemployment rates hovered around 5 percent. At the recession's worst points, they doubled.[5] But even though the U.S. economy is picking up, a complete bounce-back on jobs doesn't seem to be in the cards anytime soon. Economists at the Federal Reserve predict that unemployment will be between 7.1 and 8.4 percent in 2012.[6] Economists at the nonpartisan Congressional Budget Office are not projecting a return to a rate near 5 percent until 2016.[7] Some economists question whether a return to that 5 percent level is even possible given the changes in the way the U.S. and world economies function now, along with the layoffs resulting from reducing the size of government (federal, state, and local) to tackle deficits. The numbers can become a blur, but here's the bottom line: every percentage point of unemployment means about a million and a half Americans who want jobs but can't find them.[8]

That's an average—the jobs picture is even worse for some groups.

Even in the good old days of 5 percent unemployment, some Americans were far more likely to be out of work than others. According to the Bureau of Labor Statistics (BLS), the unemployment rate for African Americans is generally about twice that of whites.[9] People under nineteen have higher jobless rates.[10] And in good times and bad, education matters. During 2010, the unemployment figure for college graduates was only 4.7 percent. Meanwhile, the jobless rate for people without a high school diploma was 14.9 percent.[11] As the BLS itself points out, "Regardless of whether the economy is booming or contracting . . . more education is associated with less unemployment."[12]

The U.S. economy is not creating new jobs the way it used to.

The U.S. economy has had plenty of ups and downs over the past fifty years, but until the last decade, it cranked out new jobs on a pretty reliable basis. Between 1940 and 1950, the number of paid jobs in the United States grew by a whopping 38 percent.[13] Granted, part of that was World War II itself and the "Rosie the Riveter" wartime jobs, which was followed by the amazing postwar boom. But even in the 1980s and 1990s—when there were a couple of really nasty recessions—paid jobs increased by 20 percent.[14] So how has the United States done now that we've entered the twenty-first century? When the government completed its analysis of what happened between 2000 and 2010, even many experts were surprised to learn that the U.S. economy lost more jobs than it created during that period.[15] That's even worse than it sounds because there's no such thing as holding steady in this area. The various experts don't necessarily agree on exactly how many jobs we need to create to keep up with population growth, but the estimates range from about 110,000 to 150,000 a month.[16] If the economy doesn't come up with something in that range, we're going backward.

American families have been losing financial ground.
When you take inflation into account, the typical American family is bringing in less cash than it did back when "Mambo Number 5" was topping the charts* and George Clooney was still playing Dr. Ross on *ER*.[17] According to census data analyzed by the *Wall Street Journal*, the inflation-adjusted income for a typical U.S. household fell 4.8 percent between 2000 and 2009.[18] This is the first time Americans have lost ground in this way since the government began collecting the figures in the 1960s.[19] Meanwhile, the number of Americans who are poor has edged up. In 2009, more than 43 million people had incomes below the government's official poverty line of $11,161 for an individual under sixty-five and $21,954 for a family of four.[20] As the *Journal* pointed out, the economy seems to have been especially brutal for younger Americans—more than four in ten people between twenty-five and thirty-four reported incomes below the poverty cutoff.[21]

A lot of us are in jobs that are insecure and almost destined to be low-paying.
If you have special skills and work in a field that is growing like gangbusters, chances are you will do pretty well finding and keeping a job. And you'll probably be well paid too.

---

* "Mambo Number 5" also made the 100-worst-songs-ever list.

After all, employers need you, and since there aren't that many other people who can do what you can do, the boss is more likely to cough up extra dough to hire you and keep you on the job. If you don't have any specific skills—or you have skills that gazillions of other people have—you're just not going to do as well. According to a study by the McKinsey Global Institute, more than seven in ten Americans are "in jobs for which there is low demand from employers, an oversupply of eligible workers, or both."[22] These researcher types specialize in dry language, but we're sure you get the picture. To add to the gloom, the Bureau of Labor Statistics projects that the largest number of job openings through 2018 will be for cashiers, waiters and waitresses, and office clerks—positions that don't usually offer the best salaries or benefits.[23] And what happens when your job can be done by someone in another country who is more than content to take a fraction of what you've been earning? Well, your job prospects are even grimmer.

That's a formula for the rich and talented getting richer, and the rest of us muddling along. And that in fact is what has been happening for quite a while now. Over the past two decades, income for the top 10 percent of American households grew at roughly double the rate of just about everyone else.[24]*

---

* Some readers may be thinking that the major reason the rich have been getting richer is because of the lower taxes passed by the Bush administration. Taxes play a role (and we'll talk about taxes and tax policy later), but most of this difference is because these wealthy households had higher incomes, not because they paid less to the IRS.

COLLATERAL DAMAGE

These six trends are pretty bad by most Americans' lights. Lots of people are out of work, and even people with jobs are losing ground economically. But it's worse than that. Generally, we think of losing a job or having to work at a job that doesn't pay enough or offer enough security and benefits as an individual problem. If you've been there for even a short period of time, we don't have to tell you how much anguish and heartbreak it causes. But these bitter individual experiences set other problems in motion—and insidious ones at that. There's a domino effect when too many people are out of work or are scraping by in low-paying jobs with little future. Consider these:

- **Joblessness means less consumer spending.** When people don't have jobs or are in jobs that don't pay enough, they cut their spending to the bone. That means that the people who are trying to make a living selling things and offering services see their incomes drop as well. In many cases, they too have to lay off some of their workers to keep their businesses afloat. It's a fact of economic life: businesses need customers who have enough money to buy what they're selling, and when more people have jobs and money to spend, companies expand and hire more workers.
- **Joblessness leads to foreclosure and bankruptcy.** People who don't have jobs or who end up working at jobs that pay less than their old job often lose their homes and their retirement savings, or even declare bankruptcy.[25] Yes, during the mortgage bubble some people bought houses they couldn't afford, and when the economy was soaring some people just kept whipping out the plastic anytime they saw anything their little hearts desired.

They didn't save for a rainy day, and when the rains came, they went underwater. Maybe you're not feeling so sympathetic to everyone who's in bad straits these days; that's really up to you. But the truth is that foreclosures and bankruptcies spread damage far and wide. As people used to say back in the seventies, they're just not good for children and other living things. Lenders don't get their money back (and remember, not all lenders are those humongous banks and credit card companies we love to hate). Foreclosures bring down property values for people living nearby, and that includes many who borrowed wisely and have made every single payment on time. Families, neighborhoods, local merchants, and entire communities suffer too.

- **Joblessness means less money for cities, states, and the federal government.** Unless you were asleep for the past few years (and if you were, congratulations; you couldn't have picked a better time), you know that government budgets nearly everywhere have been awash in red ink. One major reason is that unemployed people and people who are earning less than they used to don't pay as much in taxes. They're not making as much money, so they pay less income tax. They buy less, so they pay less sales tax. Foreclosures and falling home prices mean local governments collect less money from property taxes. As it turns out, this is a gift of red ink that can just keep on giving. When states and cities have less tax money coming in, they have to cut their own spending, so they lay off workers. Plus they postpone building and refurbishing things, so the companies and workers who might have worked on those projects lose out too. Then there's the double whammy effect (to use the technical term): even though high unemployment rates and reduced family incomes mean tax revenues are lower at every level of government, the demand for government help actually goes up. More people need unemployment checks; more people need food stamps; more people lose their health insurance and have to go to the emergency room at public hospitals. Less money is coming in, but people—more people than before—still need these basic services that government supplies.

- **People who lose their jobs or who enter the workforce during a tough job market may never catch up economically.** Studies show that people who are unemployed for long periods of time, or young people who start out their careers during a downturn, often end up permanently losing ground. One Yale study that looked at white men who graduated from college between 1979 and 1989 found that those who graduated during the 1981–82 recession made 25 percent less their first year than those who graduated in more prosperous times. That's not so surprising. What's more striking is how much past wages can influence future wages. The study found that those graduates were making 10 percent less on average, even seventeen years later.[26]

If you sense a theme here, you're right: joblessness feeds on itself. When one person loses his job, other jobs are at risk too. If you managed to get through the last few years without taking a major financial hit, you can count yourself as one of the lucky ones. But even if you and your family escaped the worst of it, it's unlikely that you're getting off scot-free. Over time, job trends like those we've had over the last decade chip away at the vibrancy of our economy, and they eat away at the confidence and optimism of the nation as a whole. None of us can escape that.

Unhappily, turning things around is going to be tricky. Experts and economists don't necessarily see eye to eye on what's causing our jobs problems, much less how to fix them. Ask experts why the country has lost so many good jobs in recent years, and they'll point to a whole lineup of potential suspects.

- All those American manufacturers who closed plants here and opened new ones in countries where labor is cheaper.
- Government (national or local, take your pick) with its ever-expanding morass of red tape. Or Congress and the Beltway bureaucracy where the lobbyists and big campaign donors hold sway.
- Investors who decided that companies in China or Singapore or Brazil offer a better chance for growth than companies here at home. You might even be one of those yourself.
- Slick Wall Street bankers and derivative traders playing incomprehensible games with money they didn't really have and getting megabonuses for doing it.
- Learjet-loving CEOs who ran America's auto industry and other manufacturing companies into the ground.
- Labor unions that pushed manufacturing costs into the stratosphere.
- The Chinese and other foreign countries that won't play fair on their trading policies.

Or maybe it's like some diabolical magic potion—add a dash of this and a little of that, and poof! Millions of American jobs disappear.

It's a confusing and infuriating state of affairs, and frankly, a lot of us need help figuring it out. So that's our agenda for the next couple of hundred pages.

In the chapters to come, we'll run through some of the main reasons why the country has been losing jobs. We'll take a look at a whole slew of possible solutions. Since we'll all have to live with the results (or lack of them), we'll try to explain what's being proposed to spur job creation and why the people promoting these ideas believe they'll be effective. We also think it is crucial for you to understand the risks, costs, and trade-offs of the various ideas and strategies on the table. None of them comes with a money-back guarantee.

We're sure that you yourself don't have any of those refrigerator magnets that talk about "when the going gets tough, the tough get going" (or "shopping" or "boating" or any of the other popular variants). The saying is as trite as they come, but it's surprisingly apt here. The United States is a country with enormous assets, still very rich and powerful by world standards. Not to polish our own apple too much, but Americans have repeatedly shown themselves to be an enterprising and very hardworking people. We are tough. Whether we actually get going or not is now up to us.

How Low Can You Go: How Much Unemployment Do We Have to Live With?

When thinking about employment, it's worth stopping for a moment and considering the strange existence of Cosmo Kramer. Not because Jerry's neighbor on *Seinfeld* is unemployed, but because of what he tells us about who *isn't* employed.

Kramer is a man who has, to use an old-fashioned legal term, "no visible means of support." When asked what he does for money, he just gives that weird, self-assured smile and says, "Oh, I get by." At one point we find out that he's been on strike from H&H Bagels for eleven years.* He's been an author of a coffee table book (about coffee tables), a stand-in on television sets, a Marxist-influenced department store Santa, and a Calvin Klein underwear model, but those gigs don't last for long. Granted, Kramer has a certain entrepreneurial flair, but few of his ideas work out in the end.

The real point is this: how does Kramer fit into the jobs picture? He doesn't have a job. He's not looking for a job. There's almost no conceivable economic policy that could get him into a job.

The fact is that no matter how well the economy is humming or what the government does, there's a certain amount of unemployment out there. Not everybody's going to have a job. And not all of the people in that category are anything like Kramer, who seems basically content with his lot.

Economists use the term *full employment*, which basi-

---

* For many years, a famous bagel shop in New York.

cally means that everyone who wants to work and is capable of working has a job. Aiming for full employment has been official government policy for decades (the Federal Reserve's mandate, for example, is "maximum employment, stable prices, and moderate long-term interest rates"). But there's also a flip side: the *natural rate of unemployment*, which covers unemployment caused by reasons other than the usual ups and downs of the business cycle. Essentially, this means the people who could be working but aren't, for reasons that are not necessarily related to the economy.

The natural rate of unemployment includes all kinds of situations. Some people, at any given moment, just happen to be between jobs. Even in the best economies, some businesses shut down or restructure. Young people also tend to be between jobs more, as they first enter the workforce or job-hop before settling down in a career. (Economists call this *frictional unemployment*, since it's a normal part of the churn in the job market.) Then there are people who are voluntarily not working: stay-at-home parents, for example. And there are a few Kramers, wending their eccentric way through life, but nobody thinks that's a big factor here.

So what is this natural rate of unemployment? Like everything else in economics, it depends on whom you ask—and it's a moving target. In the 1990s, the conventional wisdom said the natural unemployment rate should be about 6 percent, but it ended up dropping to a little over 4 percent. For most people, that was good news, but it actually freaked out some economists, who believed it could signal galloping inflation as employers raised salaries in an effort to compete for scarce workers.[27] (A false alarm, as it turned out.)

The Congressional Budget Office, the generally accepted umpire for the government's fiscal policies, currently assumes natural unemployment should be about 5.2 percent.[28] Some experts, however, argue that recent trends in the economy may have pushed it higher. Technology could be reducing the number of workers we need to get the job done, for example. Or the Great Recession may have forced older people into early retirement when they'd rather be working. A paper issued by the Federal Reserve Bank of Cleveland argues that factors such as these have pushed the rate up to about 5.6 or 5.7 percent.[29]

So how do they measure unemployment anyway? How do you tell the difference between Kramer and the vast majority of people who are unemployed? A lot of people assume that the government determines the unemployment rate by counting the number of people who file for unemployment insurance. They do count that, but just looking at what's happening at the unemployment office would miss people whose benefits have run out or who weren't eligible for some reason (say, they quit their job voluntarily). Instead, the Bureau of Labor Statistics conducts a monthly survey covering 60,000 households, or about 110,000 people.[30] Based on the survey, people are put into one of three categories:

1. **People with jobs are employed.** That means they worked for pay either full-time or part-time during the previous week. It also includes people who were out sick or on vacation if they have a regular job.

2. **People who are jobless, looking for jobs, and available for work are unemployed.** The BLS puts some effort into defining this. You have to "have actively looked for work in the prior four weeks"

and be "currently available for work." Actively looking includes contacting potential employers and employment agencies, sending out résumés, answering advertisements, and such.

3. **People who are neither employed nor unemployed are not in the labor force.** This is the trickiest category, and the one that's most controversial. In fact, you can think of this as the Cosmo Kramer Memorial Labor Category.

Clearly some people aren't looking for jobs—maybe they're retired or have decided not to work for a while to care for children. Maybe they won the lottery or inherited a boatload of money, or maybe they're like Kramer and they just glide by somehow. However, there are people whom the BLS calls "marginally attached" to the workforce—sometimes they're called "discouraged" workers. The key factor is that they say they'd like to work, but they haven't been looking lately. The BLS describes four common reasons why people who want jobs don't look: (1) they don't think there are jobs in their line of work; (2) they haven't been able to find work, so they've stopped looking; (3) they lack needed skills or training; or (4) they believe employers are discriminating against them because of age or another reason.*

Sometimes when the economy begins to pick up after a recession, some of these discouraged workers feel less discouraged and begin looking for jobs again. That's why the

---

* The Bureau of Labor Statistics has a detailed explanation of how they determine the unemployment rate at www.bls.gov/cps/cps_htgm .htm#why, and it's worth a visit if this is a topic that interests you.

unemployment rate can rise even though more jobs are open-ing up. During the aftermath of the Great Recession, the num-ber of discouraged workers has been unusually high, because the economy's been so slow to recover that more people than usual either have given up or are sitting it out for a while. Just to give a sense of proportion, in March 2011 the BLS listed about 921,000 workers as "discouraged" because they don't think there are any jobs out there, and another 1.5 million as margin-ally attached because of school or family responsibilities.[31]

The key takeaway here is that full employment isn't the same as zero unemployment, and zero unemployment isn't possible. There's an irreducible minimum here—but that doesn't mean we shouldn't push it as close to the minimum as we can.

CHAPTER 2

HAS AMERICA LOST ITS MOJO?

What goes up must come down,
Spinning wheel got to go 'round
—*Blood, Sweat & Tears, "Spinning Wheel"* [1]

There's a familiar career path in Hollywood. A young actor rises to stardom and glides through leading roles in a string of hit movies. After some time at the top, the box office magic starts to fade. It's on to straight-to-video movies, commercials, infomercials, *Dancing with the Stars*, and reality TV. It's happened to some very talented people. Some of them climb back, and some of them don't.

Take Jack Palance. He soared into screen stardom in the 1950s and received two Oscar nominations—one of them for his role as the chillingly creepy, black-clad gunslinger in *Shane*. In the 1960s and 1970s, though, his career nosedived. At age seventy-three, in the kind of comeback story we all love, Palance pulled himself back into the limelight, winning an Oscar for his role in *City Slickers* and punctuating his acceptance speech with a set of one-handed push-ups.[2]

Just because you're good, it doesn't mean you stay on top indefinitely, and just because things go downhill, it doesn't mean you can't come back. And that pretty much sums up where the United States is right now in terms of jobs and the economy.

In many respects, we're still one of the world's best economic performers. When the World Economic Forum surveyed economists and business experts to choose which of 139 nations is the most economically competitive as of 2010, the United States came in at number four after Switzerland, Sweden, and Singapore.[3] The gurus who do the rating look at dozens of categories ranging from taxes and inflation to the threat of terrorism and employee work ethic, so this is very good news overall.

At the same time, there are signs that our position is weakening while other nations are coming up strong. The upshot is that we face some important choices in the next few years that will probably determine whether the United States continues to sit at the top of the economic pyramid, providing Americans with one of the world's best standards of living, or begins to slip downward.

A ONCE-IN-A-LIFETIME OPPORTUNITY

In the decades following World War II, the United States became the most prosperous nation in history, and the lifestyle of typical Americans—just ordinary, run-of-the-mill middle-class folk—became the envy of people around the world. We did a lot of things right after the war. We made decisions that helped strengthen and expand our economy and provide good jobs for a huge swath of the workforce, such as the GI Bill, which enabled millions to go to college, and the interstate highway system and other infrastructure investments, which helped support the economy.

But we also enjoyed a once-in-a-lifetime advantage. Europe and Japan were in ruins after years at war. Much of the rest of the world had barely entered the industrial age. Economically speaking, we had the playing field pretty much to ourselves. We worked hard, and we prospered.

But the world has been changing, and so have we. Although our economy is still a superstar by most measures, the "miraculous" post–World War II job engine is sputtering. Virtually everyone wants to rev up the country's ability to create more jobs and hold on to them, but there's not a consensus on how to do it even among economists and other experts. There's not even a genuine consensus about what the target should be—something we'll talk more about later.

DO WE HAVE TO BE THE BIGGEST
AND THE BEST TO HAVE GOOD JOBS?

Economically speaking, the United States has gotten accustomed to being the king of the hill, but as we'll discuss throughout the book, that may or may not be possible anymore. Even so, having the biggest and fastest-growing economy isn't the only strategy for improving the jobs issues that bother most Americans. We may need to figure out how to have an economy that creates and keeps a lot of good jobs here, even if we aren't winning all the speed races anymore. After all, a lot of big-time stars seem happy enough raking in big bucks appearing on TV rather than on the silver screen. And they still get to live in the Hollywood hills.

SLIP-SLIDING AWAY: TEN UNSETTLING DEVELOPMENTS
TO PONDER

Americans who don't follow economic data avidly—or who haven't traveled in China, Singapore, or Europe recently—may still think of the United States as the most advanced country in the world and as its unmatched economic powerhouse. That's still largely true. Compared to most of humanity now and throughout history, we are still doing quite nicely. But there are also signs that our decades at the top could be coming to an end. Let's take a quick look at some key changes to keep an eye on.

1. The United States will almost certainly lose its place as the world's largest economy.
The size of a country's economy is generally measured by its gross domestic product, or GDP, which is more or less the sum total of all the goods and services the country produces. For the United States, that's estimated to be about $14.7 trillion in 2010, a huge number that has so far let us hang on to the title of biggest economy in the world.[4] But China, whose economy has been red hot for the past decade or so, had a gross domestic product estimated at about $10.1 trillion in 2010.[5] Most experts expect it will beat out the U.S. economy (in sheer size at least) at some point in the next fifteen years.[6] China has already surpassed the United States and Germany to become the world's largest exporter.[7] But it's not just China that's coming up fast. If you consider the European Union as a single economy, its gross domestic product is already slightly bigger than ours.[8]

2. Our economy isn't growing as fast as it did in the past.
Between 1950 and 2000, the U.S. economy quintupled in size,[9] and that led to more jobs and a rising standard of living for millions of Americans. Lately, though, the U.S. economy hasn't been growing as rapidly, and that was true even before the Great Recession reared its unsightly head. Economists like to measure how much a country's gross domestic product is growing on a per-person or per capita basis. (Why is the per-person measure important? Share a pizza with six friends, and you'll get a nice big slice; share it with twenty, and you're going home hungry.) On that

basis, our economy grew about 2 percent a year between 1970 and 2007.[10] It may not sound like much, but it seemed to do the trick.

However, it's a human tendency to ask, "What have you done for me lately?" By that standard, the U.S. economy is starting to slip. If you look at U.S. growth more recently, between 2000 and 2007 it has been slowing, falling by nearly a quarter to about 1.6 percent a year.[11] Moreover, other economies were growing very quickly during those same years. Let's leave China aside for the moment: South Korea, Finland, Sweden, and Great Britain all outpaced us in economic growth during that period.[12] A growing economy doesn't mean that everyone in the country prospers, of course—sometimes the benefits of economic growth don't get shared broadly. It's also true that very poor countries sometimes have very high growth rates when their economies start developing because they're basically starting from the ground floor. More established or "mature" economies tend to slow down. But in general, healthy economic growth is a very good thing for people's jobs and incomes.

3. We can no longer count on having the world's best-educated workforce.

About four in ten people in the United States obtain a two-year or four-year college degree by the time they are thirty-five, and college attendance is far more widespread in the United States than in the past. So you might think that we're doing fine in this category.[13] But other countries are putting impressive efforts into educating their people, and they are pulling ahead in some respects. U.S. teens have been lagging in international tests in math and science. In the latest round, students in Australia, Belgium, Finland, Korea, New Zealand, and other countries all did better than American students.[14]

4. In some respects, our standard of living isn't the world's highest.

You can always start an argument trying to define a good standard of living or even an adequate one. For some people, it means owning a home and having a new car, and in these areas the United States does very well. Other people are more concerned about issues such as health care and retirement. One fairly neutral way to think about a country's standard of living is to look at its gross domestic product (GDP) on a per-person basis. For 2010, the United States ranked number ten, behind countries such as Qatar, Luxembourg, Norway, and Singapore.[15] Most Americans may be reasonably fine with that. After all, in winter, it's dark in Norway for eighteen hours each day,[16] and you really, really have to behave yourself if you live in Singapore. Some comparisons are also unfair: Qatar is a small country with a lot of oil and natural gas, and Luxembourg is an even smaller country that attracts lots of big banks, since it's a notorious tax haven.[17] Still, as we mentioned up top, it does depend on how you define standard of living. For Americans who believe that job benefits such as health insurance, retirement plans, and sick leave are part of that definition, the news is more mixed. Until 2010, the United States was one of the few advanced countries without a system for providing health insurance to everyone. The Obama health care legislation is a major expansion of insurance, but it's still being phased in and still being hotly debated, and while it should reduce the number of uninsured, it still won't cover everyone. We're also one of the few countries that don't require employers to offer sick leave.[18] According to the U.S. Bureau of Labor Statistics, about seven in ten workers in private industry receive health care and retirement benefits (other than Social Security); about two-thirds get sick leave.[19]

5. Government debt and deficits are at historic highs.
For thirty-one out of the past thirty-five years, the U.S. government has spent more than it has collected in taxes, and the red ink is now approaching $15 trillion. According to estimates from the Congressional Budget Office, the country will spend more than $5 trillion on interest alone between 2011 and 2020.[20] The government's own auditors call the nation's fiscal course "unsustainable."[21] The country's debt and deficit spending were a sleeper issue for many years, but in the past two years, they've become a routinely divisive topic. Even though running deficits can spur the economy in the short run (more on that in Chapter 5), it's not a good situation to be in if you're worried about keeping interest rates low and having a strong economy over the long term. The only real cure is to cut government spending and/or raise taxes, both of which may cause some people to lose their jobs. But with the government so deep in debt, we're going to be stretched making the investments in education and technology that might help the jobs situation.

6. State governments, private companies, and American consumers piled on debt too.
The federal government has been on a borrowing binge for years, but it hasn't been the only one. From 2000 to 2010, total state government and municipal debt nearly doubled, from $1.5 trillion to about $2.4 trillion.[22] The U.S. banking system lent out so much money to so many people that it had to borrow billions of dollars from the U.S. government so that the whole thing didn't collapse. Plain old citizens threw caution to the winds too. In 1980, consumer debt stood at roughly $355 billion. By 2010, it had mushroomed to $2.4 trillion, with the typical household owing some $6,500 in credit card debt alone.[23] We've been major-league borrowers at nearly every level. The Great Recession shocked a lot

of Americans—individuals, companies, and banks—into reducing their borrowing and beginning the long process of payback. It's something we have to do, but it also means that companies have less money to invest in new products and services that would lead to more jobs. And consumers don't have as much money to buy the products that are out there. We're suffering from quite a financial hangover.

7. The U.S. manufacturing sector has shrunk.

According to the CBO, the U.S. economy has lost nearly four million jobs in manufacturing since 2000, a trend that's been a double-edged sword for Americans overall.[24] Workers lose jobs, and communities can be devastated when plants close or when local manufacturers lay people off because they don't need as many workers as before. On the other hand, as the CBO points out, American consumers have benefited because manufacturing products with fewer and/or lower-paid workers makes them cheaper.[25] The CBO says there are two causes—one is that manufacturers simply don't need as many workers because of technology and other advances, and the other is that manufacturers overseas can produce the products more cheaply. In these circumstances, more and more U.S. companies have moved their manufacturing operations abroad in order to keep their prices competitive.[26] This is probably one of the toughest and most frustrating choices we face on jobs. It's one we'll come back to in later chapters.

8. Our roads, bridges, airports, and electricity grid are aging.

With all of our flashy gadgets and modern conveniences, most Americans like to think of our country as one of the most modern in the world, but we don't even make the top ten in areas such as the quality of our roads, airports, railroads, ports, electricity grid, and telephone lines.[27] We're number sixteen for broadband access and number ten for Internet access in schools.[28] At first blush, you might not think that these infrastructure issues have much to do with jobs and the economy, but being able to get people, products, and information around efficiently is crucial to having businesses that thrive (and then hire and pay well). Unfortunately, keeping up to date in these areas is massively expensive, especially in a country as large as this one. Moreover, we're already playing catch-up in comparison to some other fast-moving economies.

9. Cheap energy may be a thing of the past.

For more than fifty years, American businesses and American consumers have benefited from cheap energy. We used our healthy supplies of coal and natural gas to generate electricity affordably, and for most of that time we've had cheap gas. But our situation is changing. World demand for energy is expected to double by 2050.[29] The 2010 report from the International Energy Agency concluded that "the age of cheap oil is over."[30] The United States has only about 2.4 percent of the world's remaining oil reserves and about 3.6 percent of remaining natural gas reserves. We still have plenty of coal, but unless we invest in ways to use it more cleanly, relying on it to generate electricity pumps global-warming emissions into the air. Wind and solar power are promising, but they won't be practical solutions unless we invest billions of dollars in making them widely available and dramatically upgrade our electricity grid. Down the road, we're just not going to be able to count on energy prices staying low.

10. We're politically polarized and letting important problems fester.

If your stomach can take it, flick around the cable news channels some evening and have a look at the state of our political dialogue. Controversy is a crucial component of our political system, and it still produces sounder solutions than we would get if there were no debate. But healthy debate has morphed into bitter partisanship, which has led to gridlock on tough issues such as the federal debt, energy, immigration, and others. On some problems, we can't seem to get beyond the screaming. On others, we're reduced to relying on half measures and tiny steps while the problems keep hanging out there. The national debt keeps growing. We muddle along rather than make crucial energy decisions. Americans are divided on immigration, but no one thinks our policy is good the way it is. Everyone agrees these are solvable problems. Everyone agrees things can't go on the way they are now. But if we don't get our political act together, these problems will hasten our slide into second-tier status.

BUFFALO SPRINGFIELD AND THE WORLD ECONOMY

In economic terms, the world has been changing so fast that it's left a lot of us confused, anxious, and maybe even a little intimidated. Globalization is the challenge. (Now go find three friends who can actually define it.) How we'll fare in this new, more competitive, "globalized" environment remains to be seen. To quote those well-known international economic experts Stephen Stills and Buffalo Springfield, "something's happening here; what it is ain't exactly clear."[31]

But we've spent enough time letting confusion and economic jargon reign. It's time to reboot—we need to assess our strengths and weaknesses, consider our goals, and follow through with the best plan of attack.

So for the rest of the book, we're going to look at some of the major options out there. We'll take a look at where we are now on jobs, and we'll review some key questions on what helps and what hurts when it comes to creating and holding on to jobs here. Then we'll outline the pros and cons of some proposals likely to be on the docket as the next election rolls round. We'll even throw out a few ideas that aren't getting much attention in today's KISS (keep it simple, stupid) brand of politics. And if you feel daunted as we set out, just think about seventy-three-year-old Jack Palance collecting his Oscar and doing some one-handed push-ups to celebrate. It's on YouTube if you need a little inspiration.*

---

* www.youtube.com/watch?v=AGxL5AFzzMY.

CHAPTER 3

JUST THE FACTS, MA'AM

Ten Jobs Trends Worth Keeping an Eye On

Every day I get up and look through the *Forbes* list
of the richest people in America. If I'm not there, I
go to work.

— *Comedy writer Robert Orben*[1]

In the long-running TV series *Dragnet*, stone-faced detective Joe Friday was famous for saying "Just the facts, ma'am." There seems to be considerable wrangling among the show's fans over whether Friday ever said those precise words, but regardless of how he put it, he had no patience with embellishment. So to start this chapter, we'd like to call to your attention a specific bit of Joe Friday dialogue from the 1954 movie based on the show.

Friday is about to question a slick, well-dressed gangster who mockingly asks him, "How much do they pay you to carry that badge around? Forty cents an hour?" As always, Friday has the facts right at hand. He details his salary ($464 a month) and itemizes his deductions for taxes and his pension, along with his donation for orphans and widows. He then calculates that he brings home $1.82 an hour, and he adds, "So mister, you better settle back in that chair, because I'm going to blow about twenty bucks of it right now."[2] Of course, the cost of living was significantly lower back then, and we don't know much about where he lived or went on vacation (if he ever did). But the fact that a seemingly middle-class guy was getting by on $1.82 an hour offers a little perspective on where we've been in terms of jobs.

In this chapter, we're going on a lightning-fast tour of basic information about the American workforce: what we do, how much we earn, who has a job and who doesn't, and what the jobs picture has been like over the past few decades. We'll also touch briefly on what's happening in other countries, just for the sake of comparison.

Here are some of the key jobs facts and trends you need as background:

The number of Americans who work and want to work has almost doubled since 1970.
According to the Bureau of Labor Statistics, the number of Americans in the workforce has been rising steadily over the last few decades, and this is a very important basic number to keep in mind. In the summer of 2011, for example, there were roughly 154 million people in the country who wanted and needed jobs, compared to about 80 million in 1970.[3] This shows how crucial it is for the economy to keep creating new jobs. In many respects, the ability of the American economy to keep up with the growing number of people looking for jobs has been quite astonishing. Another way of looking at this is that the percentage of people in the labor force has stayed pretty much the same for twenty years (65.9 percent in 1988, with just a slight bump to 66 percent in 2008).[4] But we need more and more jobs (an esti-

mated 100,000 to 150,000 a month) just to keep up with the population.

Since 1970, the yearly U.S. unemployment rate has never been lower than 4 percent or higher than 10 percent (at least so far).
A lot of people initially think that in an ideal world, the unemployment rate would be zero, but having everyone employed and no one out looking for a job could cause problems. You need to have what the experts call "churn" in the workplace. People looking for jobs want there to be several openings to choose from; people hiring want to be able to look at a number of candidates.

The economic meltdown of 2008 and 2009 pushed unemployment rates up to 10 percent for a short time, an alarming jump upward, and the numbers haven't fallen as much or as rapidly as most of us hoped. The country has certainly seen high jobless rates before; in 1982, the average unemployment rate hit 9.7 percent, and in 1983, 9.6 percent.[5] But the economy bounced back and the unemployment rate fell to 7.5 percent in 1984. One of the most troubling issues facing us now is that the jobless rate seems to have gotten stuck at relatively high levels, and some experts believe it'll be staying in this vicinity for a while. Of course, nothing in the last forty years compares with what happened during the Great Depression. In 1933, the unemployment rate was 24.9 percent, and it didn't fall below 14 percent until 1941.[6]

Unemployment Rate, 1970–2010

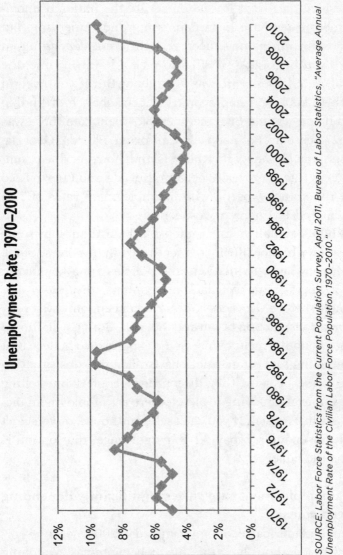

SOURCE: Labor Force Statistics from the Current Population Survey, April 2011, Bureau of Labor Statistics, "Average Annual Unemployment Rate of the Civilian Labor Force Population, 1970–2010."

The current U.S. unemployment rate, approaching 10 percent, is the highest rate in nearly thirty years.

The unemployment rate varies significantly depending on where you live.

Ever since the Great Recession set in, the national unemployment rates have been depressing and unnerving, but the circumstances have been very different depending on where you live. In April 2011, to pick a point in time, the overall U.S. jobless rate was 9 percent.[7] But if you lived in North Dakota, you had a pretty decent chance of finding a job if you wanted one; the unemployment rate there was just 3.7 percent.[8] For people in California, Florida, Georgia, Mississippi, Nevada, and Rhode Island, however, the unemployment rate was over 10 percent. Nevada and Puerto Rico were the unlucky losers, with unemployment rates of 13.6 percent and 16.1 percent, respectively.[9]

There's actually a fairly intense debate about why some states have higher unemployment rates than others. Some experts say labor regulations and policies are a key factor, while others point to a lack of economic diversity.[10] For example, before the recession, Nevada's economy was relying on a construction boom and tourism. Both got slammed in the economic meltdown.[11] Some experts also believe that plummeting real estate prices have made the jobs situation even worse because they discourage people from selling their home and moving to areas where the employment picture is a little rosier. If you can't sell your house, or you can't sell it for enough to pay off your mortgage, that option is not on the table.

The unemployment rate varies significantly depending on who you are.

When the American economy tumbled in 2008, nearly everyone worried about how safe his or her job was. Watching news footage of prosperous-looking young traders pouring out of Lehman Brothers' Manhattan offices after the firm

Unemployment Rate by State, 2010

State	Rate		State	Rate
Alabama	9.5		New Hampshire	6.1
Alaska	8.0		New Jersey	9.5
Arizona	10.0		New Mexico	8.4
Arkansas	7.9		New York	8.6
California	12.4		North Carolina	10.6
Colorado	8.9		North Dakota	3.9
Connecticut	9.1		Ohio	10.1
Delaware	8.5		Oklahoma	7.1
Florida	11.5		Oregon	10.8
Georgia	10.2		Pennsylvania	8.7
Hawaii	6.6		Rhode Island	11.6
Idaho	9.3		South Carolina	11.2
Illinois	10.3		South Dakota	4.8
Indiana	10.2		Tennessee	9.7
Iowa	6.1		Texas	8.2
Kansas	7.0		Utah	7.7
Kentucky	10.5		Vermont	6.2
Louisiana	7.5		Virginia	6.9
Maine	7.9		Washington	9.6
Maryland	7.5		West Virginia	9.1
Massachusetts	8.5		Wisconsin	8.3
Michigan	12.5		Wyoming	7.0
Minnesota	7.3			
Mississippi	10.4			
Missouri	9.6			
Montana	7.2			
Nebraska	4.7			
Nevada	14.9			

SOURCE: Bureau of Labor Statistics, "Regional and State Unemployment, Annual 2010."

went belly-up gave most of us the impression that the big jobs meltdown was going to hit pretty much everyone, even people who seemed poised for success. But once the dust cleared, it was evident that the fallout varied dramatically for different parts of the population.

Consider the unemployment rates among some different groups of Americans at the end of 2010. For college-educated white women over age forty-five, the unemployment rate was a relatively low 3.7 percent. For younger white women, between twenty-five and forty-four, who didn't finish high school, the figures were more daunting: a fear-inducing 17.5 percent. For college-educated African American men, the unemployment rate ranged from 12.7 percent for those just out of college (under twenty-four) to 6.8 percent for those over forty-five. Young men who didn't finish high school faced the bleakest prospects. For young white men without a high school diploma, the unemployment rate was more than 25 percent; for young African American men without that all-important high school degree, the jobless rate was a soul-crushing 48.5 percent.

More people are out of work for longer periods of time, and the chances of finding a job go down the longer someone has been unemployed.

Losing your job is bad enough, but the longer you stay unemployed, the worse it is. And the chance of being jobless for a long time has been historically high the last few years. By the end of 2009, more than four in ten unemployed people had been out of work for more than six months. According to the Bureau of Labor Statistics, this was "by far the highest proportion of long-term unemployment on record, with data back to 1948."[12] In March 2011, the average duration of unemployment was more than thirty-seven weeks, the longest since the Great Depression.[13]

Long-term Unemployment

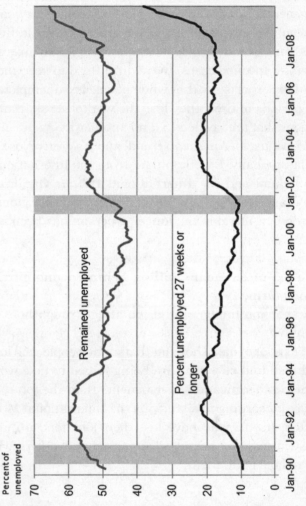

Remained unemployed

Percent unemployed 27 weeks or longer

Percent of unemployed

SOURCE: Bureau of Labor Statistics.

Not only are more people unemployed, but the number who have been unemployed for six months or longer has increased substantially. *Percentage of the unemployed who remained unemployed month to month, and percentage who were unemployed twenty-seven months or longer, three-month seasonally adjusted average. Gray shaded areas indicate recessions.*

Being unemployed for months on end is hazardous on multiple fronts. There is the danger of running out of unemployment insurance. Even before that, unemployment checks generally don't replace a worker's full salary, and people who are jobless often watch their economic lives come undone—credit card debts mount, people run through their savings. In some cases, they can't afford to keep their homes. Many experts say that when people are unemployed for long periods of time, they lose the workplace skills and connections that are needed to land another job.[14]

This isn't just an American trend, and it's pretty close to an iron law globally. Based on data from the International Monetary Fund and the International Labour Organization, the chances of finding a job anywhere in the developed world drop considerably the longer a person has been out of work.[15]

Losing a job often means falling behind economically in the long term.

Let's say that you finally land a job after a rough bout of being without one. You're back on your feet, right? Maybe not. There's a lot of data showing that many people who lose jobs and then find new ones are being forced to take work that offers lower pay and fewer benefits than the job they had before. According to a study by the International Monetary Fund, people who have lost their jobs tend to have a "persistent loss" in earnings compared to workers who manage to hold on to theirs:

Long-term Impact of Unemployment

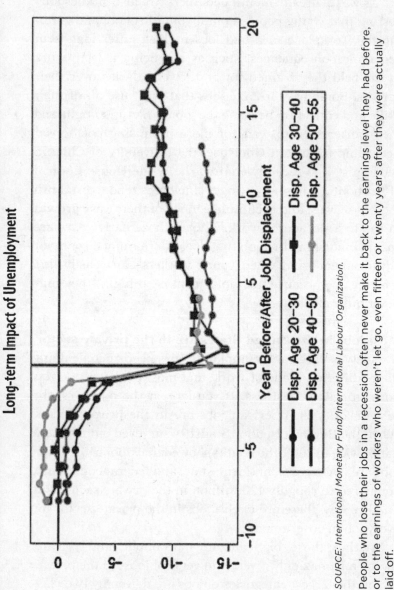

Year Before/After Job Displacement

- Disp. Age 20–30
- Disp. Age 40–50
- Disp. Age 30–40
- Disp. Age 50–55

SOURCE: International Monetary Fund/International Labour Organization.

People who lose their jobs in a recession often never make it back to the earnings level they had before, or to the earnings of workers who weren't let go, even fifteen or twenty years after they were actually laid off.

As we've already mentioned, there are also studies suggesting that young people coming out of school and entering the workplace in a bad jobs market suffer long-term economic consequences, such as not being able to work in the field they trained for and lower salaries over their careers. Some have to take jobs that don't use all of their skills and education because the job market is so tight and they can't afford to wait for more promising positions.[16] According to Austan Goolsbee, the University of Chicago economist who recently chaired the White House Council of Economic Advisers, younger employees tend to establish their base wage and accumulate much of their wage growth in the first decade of work.[17] When those early years are spent languishing in a job that doesn't offer much room for advancement—and when young workers can't easily find a better one because unemployment is so high—they may never quite catch up.

Most jobs in the United States are in the private sector. A society as large and complex as ours needs people doing many different kinds of work, and most of that work gets done in private business. If you look at the numbers, the vast majority of American jobs are in the private sector. When the Bureau of Labor Statistics surveyed employment nationwide in 2009, the results showed that about 10 million jobs are in government (local, state, and federal combined), compared to roughly 120 million in the private sector. Put another way, there are twelve jobs in the private sector for every job in government.

Within those big categories of public and private, there's an amazingly diverse universe of jobs. In the private sector, the largest categories of jobs are driven by two classic American obsessions: shopping and eating. There are more than 4 million positions in retail sales and another

3.4 million jobs as cashiers.[18] With Americans' fondness for fast food and eating out, it takes a lot of people to feed us. Whether it's White Castle, Red Lobster, or something more in the white-tablecloth-and-candlelight department, there are nearly 5 million jobs for people to prepare food and serve it to others.[19]

On the public sector side of things, there's more variety here than you might expect. We sometimes think of government workers as "paper pushers" and "faceless bureaucrats," but nearly a quarter of the jobs in the federal government are connected to the post office.[20] We guess you could say the workers are pushing paper, but it's mainly sorting out letters and flyers and catalogs and putting them into our mailboxes. Jobs in state government run the gamut, but more than 250,000 of them, about 6 percent of state government workers nationwide, are corrections officers and jailers.[21] In local government, education and police work are the big categories. About 20 percent of local government workers are teachers.[22]

Employment by Industry, 2007

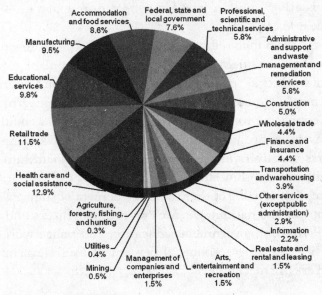

Note: Total number of jobs: 130,647,600

SOURCE: *Bureau of Labor Statistics, "Occupational Employment and Wages, Annual 2010."*

There's an amazing diversity of work out there, and almost all of it is in the private sector.

There are more low-wage jobs than high-wage ones, and Americans' wages over the past decade have barely kept up with inflation.

Many jobs in today's economy don't pay all that well. Jobs as salespeople, cashiers, and waiters and waitresses are a big slice of the employment pie, and these are typically lower-paying jobs that don't require college degrees or much experience (snooty sommeliers in fancy-shmancy restaurants notwithstanding). According to the Bureau of Labor Statistics, "most of the largest occupations were relatively low paying." The BLS calculates the average salary in the United States at $21.35 an hour, coming out to $44,410 annually.[23] But the typical pay for cashiers is $9.52 an hour, while the BLS category of "combined food preparation and serving workers" pays $8.75.

Of course, some occupations are much more lucrative, but most of us probably couldn't do these kinds of jobs even if there were more of them available. The BLS calculates that there are about 45,000 jobs for surgeons, and on average they earn about $108 an hour.[24] The country's 13,250 economists earn an average of more than $47 an hour. (We'll let you decide for yourself whether economists are worth that much money or not.)[25]

On average, workers' wages haven't grown much over the last decade once they are adjusted for inflation. The BLS tracks the typical weekly salary for full-time wage and salary workers in inflation-adjusted dollars. In 2001, this typical weekly wage was $336. In 2010, it was $343.[26] But here again, you need to keep the huge variations in salary in mind. In 2010, for example, the typical weekly wage for people with jobs in "management, professional, and related occupations" was more than $1,000.[27]

Median Weekly Earnings, 2002–10

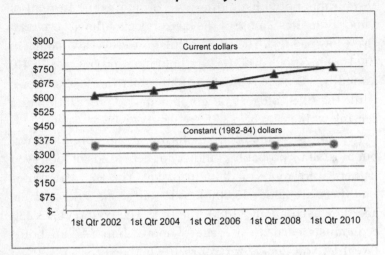

SOURCE: Bureau of Labor Statistics, "Occupational Employment and Wages, Annual 2010."

Median earnings for American workers have stayed flat for the past decade when adjusted for inflation.

Many experts see the gap in both jobs and wages continuing or even increasing. Some recent analyses suggest that new jobs coming online tend to be either well-paying jobs requiring top-notch skills and education or low-wage jobs demanding minimal training—it's the jobs in the middle that are disappearing.[28]

Surveys show that most Americans are concerned about the growing income gap between the wealthiest Americans and everyone else, and of course, these huge variations in what people earn play into that. It's a topic we take up in more detail in Chapter 11.

The economy's ability to create jobs has also been lagging.
As we'll discuss in later chapters, figuring out how to spur
job creation—not just prevent layoffs—is vital to turning
the country's unemployment picture around. And job creation, especially in the private sector, has been slipping for
the last decade.

Private Sector Job Growth Lagging

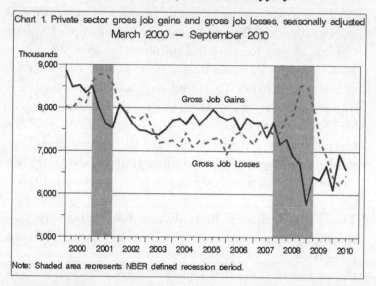

Chart 1. Private sector gross job gains and gross job losses, seasonally adjusted March 2000 — September 2010

Note: Shaded area represents NBER defined recession period.

SOURCE: U.S. Bureau of Labor Statistics.

The private sector provides most of the jobs in the United States,
but private job creation has been sliding for the past decade.

As economist and author Michael Mandel has pointed
out, the picture is even more troubling because most job
growth over the past decade or so has been in government jobs—not in the much larger private sector. As we
mentioned, there are twelve private sector jobs for every
government job, so sluggish private sector job creation is
bad for the economy overall, no matter who signs your

paycheck. What's more, with budget cutbacks in federal, state, and local governments, the economy is going to need even more private sector jobs to make up the difference. In the last decade, government employment rose by about 2.4 million jobs, while the private sector only added about half of that—1.1 million jobs between 1999 and 2009.[29] What's more, the pace of creation varied significantly depending on what kind of work you're talking about. Reasonable numbers of jobs were created in health care and education (where there's a lot of government spending) and the food service business (not a big government area, but we do like our burgers and fries in this country). At the same time, there were big job losses in manufacturing. All that suggests that government has been doing more than its share to keep the job machine going, while the private sector sputters.

The United States has always had relatively low unemployment compared to other industrialized countries.

Compared to some developing countries, the unemployment rate in the United States seems magical. The poorest countries in Africa, for example, have unemployment rates over 40 or 50 percent—50 percent in Namibia, for example, and 48 percent in Senegal.[30] But even compared to countries with economies that are more similar to ours, the United States has done reasonably well. At least, we have in the past.

The "who does it better" rivalry between Europeans and Americans will probably continue until we've all gone to our final reward (whatever that may be). In the jobs category, however, the United States has long

had a reasonably good track record of keeping unemployment at manageable levels compared to European
nations, at least until the Great Recession knocked
us back on our heels. In fact, this country's success
at keeping joblessness relatively low prompted many
European experts to urge their countries to adopt some
of our policies, especially those giving employers more
flexibility in hiring.[31]

International Unemployment Rates, 2010

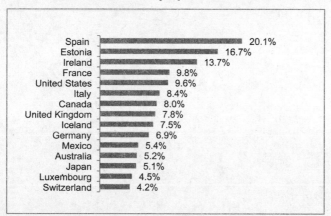

Spain	20.1%
Estonia	16.7%
Ireland	13.7%
France	9.8%
United States	9.6%
Italy	8.4%
Canada	8.0%
United Kingdom	7.8%
Iceland	7.5%
Germany	6.9%
Mexico	5.4%
Australia	5.2%
Japan	5.1%
Luxembourg	4.5%
Switzerland	4.2%

SOURCE: Organisation for Economic Co-operation and Development, February 2011.

It could be worse, but it could be better too. There are countries with persistently higher unemployment rates than the United States.

This country's comparative advantage is particularly
striking when it comes to having jobs for younger people.
The group of those under twenty-five has had a rough
time in the current U.S. job market, no doubt about it,
but joblessness among young Americans is much less

severe than it is for young Spaniards or young Italians.[32]

Lest we get too impressed with ourselves, though, we'll be taking a look at some areas where the United States is lagging behind some other countries. Education is a particularly sobering example, and education levels could be pivotal to the country's ability to pull itself out of its current economic predicament. We'll be taking up this question in Chapter 9.

Roller Coaster or Dark Ride?

Is Today's Joblessness a Temporary Setback, or the Start of Something Different?

We all know something's wrong with the way our economy produces jobs—if you didn't think so, you would have set this book aside by now. We generally try to avoid the wonky terms economists use, but there's one concept that gets a lot of discussion, and it's worth understanding if you're going to follow along with the jobs debate in the media and political circles: *cyclical* versus *structural* unemployment. In a noneconomist's terms, the basic question is whether the stubbornly high unemployment rate means that there aren't enough jobs to go around because of the Great Recession, or whether it means that the workers we have don't fit the jobs that are out there.

The New Normal

Cyclical unemployment happens in every recession. Let's say, for example, that a nice town called Shelbyville has an auto plant. When a recession hits, people don't buy as many cars, so there's no need to make more of them. The automaker cuts back and lays off workers, or even closes the plant and puts everybody out of work. The good news, however, is that when the economy picks up again, the automaker will bring workers back. Or, if the auto jobs don't return, at least a prosperous economy will generate other jobs in town that those unemployed autoworkers could get.

But in the structural model, suppose that the autoworkers aren't qualified for the new jobs that come into town. Maybe the people in Shelbyville are aging, and the new growth industry is in health care and nursing homes. An autoworker who knows how to assemble an engine doesn't know how to put in a catheter (and it's not something you want to learn on the job). So the autoworkers are going to stay unemployed, unless and until they retrain for the new jobs.

Alternatively, let's say that even though the auto company closed its Shelbyville factory, it is hiring at its other plants in nearby states. Maybe the laid-off auto workers would be better off moving to Alabama and taking jobs in the automaker's plant there—that is, if they are willing and can afford to move. If the Shelbyville housing market has tanked, those autoworkers may not be able to sell their houses and move someplace where the job prospects are better. Moreover, if the workers have kids in school, a spouse who works, or elderly parents nearby, moving to take a new job can be more complicated. Sometimes people don't know that there are jobs in other places, plus moving is costly in and of itself. Meanwhile, the Shelbyville nursing home may actually have jobs going begging, because there aren't enough nurses and trained health aides in the community.

That's what structural unemployment looks like, and one of the fiercest debates in economics over the last couple of years is whether the American economy is facing a mismatch between unemployed workers and the jobs that are out there.

"It's a Mismatch"

The president of the Minneapolis branch of the Federal Reserve, Narayana Kocherlakota, caused a major stir when he pointed out that job openings increased 20 percent between June 2009 and June 2010, but so did unemployment. He speculated that as much as 3 percent of the 9.5 percent unemployment rate in 2010 could be linked to structural factors. "What does this change in the relationship between job openings and unemployment connote?" Kocherlakota asked in a 2010 speech. "In a word, mismatch. Firms have jobs, but can't find appropriate workers. The workers want to work, but can't find appropriate jobs. There are many possible sources of mismatch—geography, skills, demography—and they are probably all at work."[33]

Experts who believe structural forces are at work point to the fact that people with high school educations are unemployed more often and longer than college graduates, and that areas with high unemployment are also the places where large numbers of people are "underwater" on their mortgages—they owe the bank more than their house is worth. They also argue that since one of the main reasons for the Great Recession was a collapse of the housing bubble, lots of construction workers are out of work and untrained for other jobs.

Other economists push back on this. While it's true some industries and less-educated workers have suffered more than others, the Great Recession hit almost every sector and state to some degree. Plus, when businesses have jobs but can't find enough workers to fill them, they generally offer higher wages. This would certainly be something you would expect businesses to do if they needed to persuade workers living in one place to move and come to work for them, but there's not much evidence that this has been happening during the recession.[34] According to an analysis by Payscale. com, wages at the end of 2010 were about where they were at the beginning of 2008, almost three years earlier.[35]

It's the Implications, Stupid

There are two things you should keep in mind as you listen to the argument over whether our current jobs problems are cyclical or structural. One reason this debate has split the economics world into camps has as much to do with the implications as with the evidence. It's political, frankly. If the main cause of our current unemployment woes is cyclical—caused by a temporary downturn in the economy, not some major mismatch between jobs and workers—then the solution is to push the overall economy forward by tax cuts, more government spending, or other stimulus efforts.

If the problem is structural, however, then those efforts won't do much good. In some cases, the economy works out those structural problems reasonably well on its own, as workers either retrain themselves or move to someplace with more opportunity. But the government may be able to help with job training programs, or with targeted tax incentives directed to specific communities or industries. For example, maybe Shelbyville needs to attract some different types of manufacturing companies to the region so that the automakers' skills could be upgraded with modest on-the-job training. That's different from trying to help an auto plant manager become a registered nurse.

But some structural mismatches are incredibly complicated, and the solutions are a lot trickier. What about jobs that are eliminated by big-picture trends in technology, for example? What about the fact that many of those manufacturing jobs used to be good jobs for high school graduates without further training? Not only is there a mismatch between jobs and workers, there may be a mismatch between jobs and the education system. Plus, if technology allows manufacturing to make just as many products with fewer people, some jobs are gone forever. They won't come back simply because the overall American economy is pulling out of a recession.

In the current, ham-handed way our political system handles issues, this is being defined as a debate about doing something versus doing nothing. That's wrong. In reality, what we're facing is a question of how we attack both types of unemployment, the cyclical and the structural. We need to leave the aftereffects of the recession behind us, and we need to think about the future of jobs—or we could end up losing on both fronts.

CHAPTER 4

CHURN, BABY, CHURN

Why Creating New Jobs Matters So Much

Please release me, can't you see
You'd be a fool to cling to me.

—*Elvis Presley, "Release Me"*[1]

Here's a story about Elvis.

In 1954, he was working at the Parker Machinist's Shop in Memphis and hoping to become a singer. He recorded a few songs at a local studio, the now legendary Sun Studios, and one of them, "That's All Right," was played by disc jockey Dewey Phillips on his *Red, Hot, and Blue* radio show.[2] When dozens of thrilled young listeners called in, the DJ played the song fourteen times. Elvis was on his way.[3]

So what does this have to do with America's jobs problems? Think about the disc jockey for a moment. Here's a job that didn't exist—and which no one would have thought of—before the invention of records and radio. In the 1950s and 1960s, those technologies merged with a rising youth culture and sophisticated marketing that made DJs such as Wolfman Jack household names. Even a relatively obscure

DJ could help usher an unknown local singer into stardom, driving teens to buy records at the music stores that existed in nearly every town.

ELVIS GONE VIRAL

That's not how the music business operates today. People buy CDs, but you're just as likely to download music onto your iPod. The local music store is a vanishing phenomenon. Even big chain stores such as Tower Records and Virgin Megastores have closed up shop in the United States. There are still people who earn their keep choosing music for the rest of us to listen to, but nowadays Elvis could record his own songs and upload them to YouTube. He'd have a website and Facebook page. We'd gauge his popularity by the number of people following him on Facebook and Twitter. We would be far more likely to learn about Elvis from a video gone viral than from a DJ. Disc jockeys aren't really where it's at in music anymore.

THE GREAT DISAPPEARING ACT

The point is that in a healthy economy, jobs appear and disappear all the time (the experts call it *churn*). As changes in the music business mean fewer jobs for disc jockeys and at brick-and-mortar music stores, new jobs are being created at online music outlets like iTunes and Rhapsody. There is a creative, changing process at work.

In a prosperous economy, the number of jobs being created outpaces those being lost, and the fact that some jobs naturally fade away doesn't attract that much attention. When unemployment rises, we all get preoccupied with how many jobs have been lost. But how many new jobs are created is just as important as how many old jobs are going away—and maybe more so.

Many experts have laid out the math on jobs in the wake of the 2008 and 2009 economic nosedive. Between December 2007 and February 2010, the economy lost an estimated 8.4 million jobs, or "more than 6 percent of the prerecession job base," as economist Mark Zandi of Moody's Analytics puts it.[4] During that time, the country would have needed to add between 110,000 and 150,000 jobs a month just to keep pace with population growth and people entering the labor market.[5] And remember our Joe Friday "just the facts" data from Chapter 3. From 1999 to 2009, the private sector created just 1.1 million new jobs *in the whole decade.** We need about that many new jobs every year just to keep up.

So what happens when you run the numbers? In 2009, the *Wall Street Journal* calculated that it would take eighty-six months—just over seven years—for the economy to get back to prerecession levels.[6] In 2011, the Congressional Budget Office projected unemployment wouldn't fall back to 5.3 percent (roughly full employment) until 2016—seven years after the official end of the recession.[7] Talk about seven years of famine. That's pretty much what we're facing on jobs.

JUST STOP THE BLEEDING

Typically, whenever the economy goes into recession, the government's response is much like what happens in an emergency room after someone's been shot: you stem the bleeding and treat the shock first. You don't worry about

---

* Government jobs supplied another 2.4 million openings, but that's unlikely to continue given the serious budget problems at the federal, state, and local levels, especially when many Americans believe that government should be smaller, not adding jobs.

whether the patient was healthy before the bullet hit. The federal government certainly tried to mitigate the damage in 2008 and 2009, when the economy took a nosedive. To quote the inimitable Frankie Valli, sometimes it's a good idea to just "hang on to what we've got."

It may be worth revisiting exactly what the government did on jobs in the $787 billion stimulus requested by President Obama and approved by Congress in 2009.* In essence, the government did try to preserve existing jobs, but since you can't preserve a job that's already gone, the stimulus also attempted to create new ones in hard-hit industries. The legislation basically used three strategies:

- It included about $140 billion for the states and cities to help them with their own budgets so they wouldn't have to lay off workers such as teachers and police officers during the economic crisis.[8]
- It cut some taxes for individuals and businesses, which can both save existing jobs and spur job creation.[9] When people have more money in their pockets, they're more likely to spend, so that helps businesses avoid layoffs. Lower taxes for businesses can also help existing companies stay profitable so they don't eliminate jobs, and they can encourage entrepreneurs to start new enterprises, which creates new jobs.

---

* People often treat the stimulus and the Wall Street bailout as if they were the same thing, but they're not. The stimulus was direct government spending and tax cuts designed to get the economy going. The bailout entailed $700 billion in loans to banks, brokerages, automakers, and insurance companies to keep them from collapsing in the global financial crisis. If those businesses had gone under, that would have cost jobs too, and not just at those firms. The other key point about the bailout is that the companies have paid almost all of those loans back, so the government nearly broke even on the deal.

- Finally, it provided money to build things such as highways, high-speed rail, green energy projects, and the like. Because of the collapse of the housing market, construction jobs were particularly hard hit, and some experts say that encouraging new building projects is especially important.[10]

Fewer than three in ten Americans think the administration's stimulus plan helped the economy,[11] and experts too are sharply divided on whether the stimulus was the best way to respond to the economic crisis and whether the government should have spent more or less money on it. The Obama administration didn't do itself any favors when it predicted the stimulus plan would keep the overall unemployment rate at 8 percent; instead it hovered near the 10 percent mark for months.

But if you're someone who believes that government should act to save jobs when the chips are down, it is hard to argue that the stimulus was a total failure. According to the nonpartisan Congressional Budget Office, it generated jobs for somewhere between 1.4 million and 3.6 million more workers and reduced the unemployment rate between 0.8 percent and 2 percent in the second quarter of 2010 alone.[12]

In the worst recession since the Great Depression, numbers like that mean that millions of real human beings were considerably better off than they would have been without government help. But there were also downsides. The government spent money it really didn't have on the stimulus, and the plan's impact dwindled once the federal money ran out.[13] Even if you believe that the jobs saved were worth the money spent to save them, the federal government can't afford to prop up state budgets for too long. In general, state and cities need to be able to support themselves. There's also considerable debate among economists over whether the other parts of the stimulus—the tax cuts and money for building projects—were successful weapons against the recession.[14]

SEE THE USA IN A CHEVROLET

The federal government also tried to save jobs by offering a $64 billion financing package to General Motors and Chrysler when they were catapulting toward bankruptcy in 2009.[15] Founded in 1908, headquartered in Detroit, employing more than 205,000 people, and doing business in more than 120 countries,[16] General Motors had been losing money for several years even before the economy tanked.[17]

So the government had a choice, and if you think back to 2009, it was a pretty grim one: let GM and Chrysler go under and watch a million or more jobs disappear right when the U.S. economy was already teetering,* or use taxpayer money to salvage two companies that richly deserved to fail—at least according to many experts and much of the public. The government had already bailed out Chrysler back in 1979. All three major U.S. automakers had been struggling for years since Japanese car companies began showing how nimble and efficient they could be.[18]

The auto bailout was and still is bitterly controversial.

---

* Estimates of how many jobs would have been lost if General Motors and Chrysler went bankrupt vary. At the time, the Center for Automotive Research, an industry group, suggested that 2.5 million jobs would be lost if General Motors closed down, including jobs in related industries and retailing that are tied to the sale of GM cars. Moody's Mark Zandi believed that estimate was high. He thought the loss would be more along the lines of 1 million jobs in the first quarter of 2009. Quoted in Michael McKee, "GM, Chrysler Failure Would Push Economy in Abyss," Bloomberg, December 15, 2008, www .bloomberg.com/apps/news?pid=newsarchive&sid=agI3NWvegYUI.

President Obama and others argued that saving jobs in the auto industry—and ensuring that the United States still had an auto industry—justified bold government action, especially with the economy already so weak. Others saw the bailout as unnecessary and even counterproductive. Economist Martin Feldstein argued that letting the companies go bankrupt would have been the far wiser option. "If the companies file under Chapter 11," Feldstein wrote at the time, "they would be able to continue producing cars, and the workforce would remain employed while the firms reorganized. . . . The bankruptcy court could require the unions to rewrite contracts, bringing wages down to levels that would allow the firms to compete and, therefore, to maintain employment."[19]

WHICH JOBS HAVE TO GO?

Even worse, many critics argued, the auto bailout put the government in the position of trying to manage a private manufacturing company, which is not necessarily its strong suit. With taxpayer dollars on the line, the Obama administration naturally wanted GE and Chrysler to operate more efficiently. It pulled together a team of experts that pushed the companies to, among other steps, sharply reduce the number of car dealerships in communities across the country.[20] But car dealerships employ people too. According to the government's own inspector general, "tens of thousands of dealership jobs were immediately put in jeopardy . . . in the face of the worst unemployment crisis in a generation."[21] Rescuing the two car companies may have saved some jobs, but the government's decisions afterward also endangered others. Was it worth it or not?

For President Obama, the answer is clearly yes. Speaking to a group of cheering workers at a Jeep Grand Cherokee plant in 2010, the president summarized his point of view: "They said we should just walk away and let those jobs go. Today, this industry is growing stronger. It's creating new jobs."[22] General Motors posted a $4.7 billion profit in 2010, its first profitable year since 2004.[23] It's still an open question whether U.S. carmakers can produce the kinds of automobiles that large numbers of customers here and abroad will buy over the long haul.

So here are two examples of the government trying to save jobs. One was unpopular but arguably successful, at least in the short run. The other? Well, the results are mixed, and truthfully, it may be years before we really know whether leaping in to bail out Chrysler and GM will ultimately pay off.

What both examples do show is that trying to save jobs is a contentious, messy, and complicated business. For politicians in the midst of a major economic crisis, it probably falls into the category of "damned if you do and damned if you don't."

THE EVE OF DESTRUCTION?

There is a continuing battle among the experts over whether government should intervene to save jobs in a crisis, and an even bigger one over what counts as a bona fide crisis. Talking about the long term, however, there's not much disagreement at all. Focusing just on saving jobs is a terrible idea.

Economists often point out that healthy economies need and depend on "creative destruction" to make progress. There is a natural life-and-death cycle for companies and jobs, in which failing enterprises die off and healthy new ones arise to take their place. Messing with that cycle is a little like messing with Mother Nature. In fact, Joseph Schumpeter, the economist who developed the concept and coined the term *creative destruction*, compared the process to evolution.[24]

With so many Americans looking for work, the idea that it's better to let companies fail and watch their employees lose their jobs may sound heartless, but companies actually die off all the time. In their entry on creative destruction for *The Concise Encyclopedia of Economics*, Michael Cox and Richard Alm recounted the "pattern of destruction and rebirth" in major American companies over the last century. Their results: "Only five of today's hundred largest public companies were among the top hundred in 1917. Half of the top hundred of 1970 had been replaced in the rankings by 2000."[25] In June 2009, for example, the Dow Jones Industrial Average—a widely accepted list of the country's most important companies—dropped the nearly bankrupt General Motors and replaced it with Cisco, the computer networking giant.[26]

The problem is that trying to save existing jobs at all costs can introduce too many distortions into the system. Sooner or later, companies end up with more workers than they need, which sooner or later pulls them under. The pressure to save jobs can come from government, unions, or sometimes management itself, but in most cases it just postpones the inevitable. And unfortunately, the time and money spent trying to hang on to existing jobs can give your competitors—other companies, other products, other countries in the international marketplace—the chance to move in and take your place.

WE'D LIKE SOME MORE, PLEASE

So enough of Professor Schumpeter for now, because our key message here is that just saving the jobs we currently have won't do the trick on its own. We're way behind, and

we need to catch up. The United States needs more jobs created, enough to replace the more than 8 million lost in the
recession and provide work for the country's growing labor
force. This is basically what the rest of the book is about. In
Chapters 5 through 11, we zero in on different ideas about
what the country can do to rev up job creation.

Before discussing the options, though, we want to touch
briefly on how new jobs get created, which is actually quite
straightforward. Jobs are created when there is work that
someone needs done and when the boss is willing to pay
for it. Here are some examples—many of them are quite
familiar:

- **Someone starts a small business.** Once it's up and running, the owner hires people to help with the work. Think
  of any small, owner-operated business you patronize—
  hair salon, coffee shop, small clothing store, whatever.
  In the United States, there are about 3.7 million firms
  employing fewer than ten people each.[27] Most of the owners might like having another hand on deck if they could
  afford it and if they were convinced that it would help
  them make more money. Let's say all these small business owners hired one more person: the country would
  have another 3.7 million jobs.
- **The small business proves to be very popular and
  expands.** Maybe it opens branches around the city.
  Maybe it expands nationwide or even internationally. In
  1971, Starbucks was a small coffee shop in Seattle's Pike
  Place Market, and it stayed that way for about a decade.
  In the 1980s, Howard Schultz became the head of the
  company. (He actually started off as a customer, but he
  had this bee in his bonnet about coffee shops and how
  every community should have one.) Now there are fifteen thousand Starbucks in some fifty countries.[28] And

the jobs? Serving the lattes and mocha frappuccinos in the United States alone provides work for nearly 140,000 employees.[29] Plus the coffee shops have to be built, decorated in the company's very own style, and filled with products ranging from coffee machines to all manner of merchandise flaunting the Starbucks logo. All this stuff has to be produced and shipped around the country. It's a lot of jobs.

- **Someone comes up with a useful product.** The big idea that sets jobs creation in motion doesn't have to be earth-shattering. It only needs to be a product or service that's useful and that people want. In the early 1950s, Alex Manoogian started manufacturing a single-handle washerless faucet. You know how older bathrooms have separate faucets for hot and cold water and you have to mix the water yourself to make it warm? A trio of tinkers had come up with something called a ball-valve joint, which allowed hot and cold water to come through the same faucet. "It appealed to me," Mr. Manoogian said later. "I said, 'Why wouldn't people want to do something with one hand instead of with two?'"[30] That ended up launching Delta Faucet and its parent company, the Masco Corporation.[31] Masco, still headed by the Manoogian family, now employs some forty thousand people and makes all sorts of products in the United States and Europe, including "faucets, kitchen and bath cabinets, paints and stains, bath and shower units, spas, showering and plumbing specialties, windows and decorative hardware."[32] Think about that the next time you're standing in front of the sink adjusting the water to exactly the temperature you want.

- **Someone invents something entirely new, and a whole industry grows up around it.** Consider Steve Jobs and his less famous but equally enterprising partner

Steve Wozniak. When they were in their twenties, they scrounged up $1,300 to make computers in Steve Jobs's garage. The new Apple Computer's first big break? They sold fifty computers to a local Bay Area store.[33] Now the company is headquartered in Cupertino, California, on a street named Infinite Loop. What with the iPods, iPads, iPhones, iTunes, and Mac Books, the company's future looks infinite too—at least for now. Apple itself employs just under ten thousand people, but that's only the beginning. Apple needs stores, marketing, advertising, shipping, and much more to get its products into the hands of customers. In 2010, the larger family of Apple employees was "approximately 46,600 full-time equivalent employees and an additional 2,800 full-time equivalent temporary employees and contractors."[34] Even more important, there are now hundreds of companies and jobs that simply wouldn't exist without modern computing, and modern computing advanced on the backs of Apple and, of course, Microsoft. These companies started small, but their success led to millions of jobs.

- **People's lives change.** Jobs can also be created when people start to want and need products and services that they didn't need before. In the 1950s, when most mothers raised their children at home, there wasn't much of an organized child care industry. Now, with some six in ten mothers of children under eighteen working outside the home—and stay-at-home dads still in short supply— there's a whole new system of day care centers and services providing nannies and babysitters. According to the Bureau of Labor Statistics, about 1.3 million people in the United States work in some form of child care. The BLS estimates that about a third of these workers are self-employed.[35] Less happily, we now have companies providing "nanny cams" so parents can reassure themselves that their worst fears are never realized. And with everyone in the family so busy, companies that supply takeout food, fast food, prewashed lettuce, online food shopping, and delivery pizza are in their glory days. As we mentioned in Chapter 3, there are lots of jobs in food in all its guises.

- **Government sets something in motion.** Most jobs in the United States are created in the private sector, but government can spur job creation too, and not just by adding people to its payrolls (although that is one method). In the 1970s, Congress passed legislation requiring public schools to teach children with special needs: those with physical or learning disabilities. (Before then, public school systems didn't have to accept special-needs children; they'd be sent off to special schools, if there was room, or just ignored.) By 2008, according to the BLS, there were 473,000 special education teachers in the United States,[36] and the Bureau projects that future employment in special education will "increase faster than the average for all occupations." As the Bureau puts it: "Job prospects should be excellent because many districts report problems finding adequate numbers of licensed special education teachers."[37]

HOW ARE JOBS CREATED? LET US COUNT THE WAYS

Our point is that jobs come into being in innumerable ways, and given the scale of the challenge the United States faces now, we need to be creative, open-minded, and thoughtful about how to make more of that happen. Companies come in all sorts of shapes and sizes with as many purposes as the human mind can devise. Government hires people at the local, state, and national levels. Those jobs count too.

And there are nonprofit enterprises that hire people. We work for one of them.

The decisions we make as a people—through government—can set the stage for job creation or stifle it. Then there is what we all do ourselves—what we buy, where we live, how well we educate ourselves for our jobs, whether we summon the get-up-and-go to look hard for work or decide to start up something for ourselves. According to the Department of Labor, more than ten million Americans are self-employed and another million or so get additional income from some form of self-employment.[38] If you can make it being your own boss, you're creating a job as well.

So that's the bottom line for us. Jobs are created in many, many ways. The question we face now is how to encourage as many new jobs as possible.

THINK LIKE AN EMPLOYER

When economists start talking about the "supply side" and the "demand side" and "productivity" and the "business cycle frequency," it's tempting to stand back and let the folks with the spiffy verbiage sort out what's needed to spur job creation. Maybe this is just too tough for us regular folk.

But there are important reasons you should (and can) think this issue out for yourself. One is that economists themselves don't agree on what will promote more hiring. There are different points of view with starkly different implications. Second, we will all have to live with the results, so maybe this is not something we want to delegate to a select few, no matter how smart they seem. Third, trying to determine what works to create jobs often boils down to trying to figure out human nature: in this case, how human beings will react given different kinds of incentives and realities. In this arena, average folk sometimes do better than the economists, even with all their degrees and computerized economic models. Don't sell your own judgment short.

But it's not enough to think only about what you personally would like to happen. Since we're talking about creating jobs and hiring people, we need to consider the problem from the employer's point of view. Here are some facts and ideas to chew on.

Start-up businesses are doing most of the hiring. If you think about which companies (and government and nonprofit enterprises too) are most likely to need more employees, it's common sense that those that are just getting off the ground and starting to grow are the most likely to create new jobs. And there's compelling research from the Ewing Marion Kauffman Foundation showing just that: most *new* private sector jobs are coming from businesses that are between one and five years old. According to the Kauffman report, the job creation prowess of start-ups is "dwarfing" the jobs creation record of more established enterprises.[39]

Starting a business is risky. It's logical that a newly opened business would need more help, but the owners have to be careful. Hiring people is not their top priority. Their top priority has to be to build a customer base and get their business to be profitable so they don't lose the money they've put into it. According to Kauffman's Robert Litan, half of start-up firms don't survive five years.[40] So new business owners—all business owners, in fact—have to think about hiring people in a larger context. Even if they could use some extra help, it's not a good decision for them to hire unless they believe it will pay off in better performance and profits.

It's not just the salary. Courtesy of AccountingCoach. com,[41] here's a list of expenses employers have to pay for each person they hire on top of paying the salary: Social Security and Medicare taxes (both employers and workers pay these taxes), state and federal unemployment taxes, and workers' compensation insurance in case the employee is injured on the job. Employers with more than fifty employees will face penalties beginning in 2014 if they don't offer health insurance.[42] Many employers also provide life and disability insurance and dental and vision coverage, and some contribute to pensions or 401(k) plans for each employee. The actual costs vary from state to state and industry

New Jobs Created by New Firms

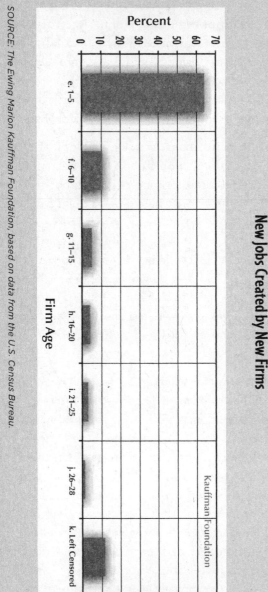

Firm Age

SOURCE: The Ewing Marion Kauffman Foundation, based on data from the U.S. Census Bureau.

Which companies create the most new jobs? Companies that are five years old or less. It's the job creation power of start-ups.

to industry, but one rule of thumb is that an employee's full cost can be 25 percent to 40 percent more than base salary. In other words, an employee who's making $50,000 per year really costs the employer up to $70,000.[43]

And even vacation time and paid sick leave cost employers money. (Remember Ebenezer Scrooge complaining about paying Bob Cratchit for Christmas Day when he was going to be home with Tiny Tim and the family and not in the office?) Employers have to be mindful of how the work is going to get done, so this affects how many people they need on the job overall. Many employers see these additional obligations as their responsibility and a good investment that helps them keep talented, industrious people working for them and not for someone else. At the end of the day, however, this is extra cash going out the door, and it has to affect the business's decisions on whether to hire or not.

Can the employee really do the job? Employers need more than warm bodies. They need people who can and will do the work that needs to get done. For some jobs, working hard, showing up on time every day, and being willing to learn on the job are enough. Sad to say, not everyone fulfills his or her part of even this basic a bargain once hired. But 6 in 10 employers also say that they have a hard time finding workers with the skills they need—especially specialized or technical skills.[44] When the Bureau of Labor Statistics compiled a list of the 30 fastest growing occupations between now and 2018, jobs like biomedical engineers, network systems analysts, financial examiners, medical scientists, biochemists, and biophysicists placed in the top 10.[45] When employers can't find the kind of help they need in the United States (skilled or not), they may outsource the jobs overseas or bring in workers from abroad. Some employers need unskilled labor like garment workers, but major employers often seek special visas to be able to hire highly skilled people from abroad.[46]

It has to make sense for the bottom line. For anyone running a business, even a nonprofit one, the reason to hire someone is not to give them a job—it's to get the organization's work done as efficiently and cost-effectively as possible. When technology allows an employer to get the job done more easily with fewer people, that's what nearly all of them will do—what they'll *have* to do eventually. The same goes for moving jobs to other countries or hiring temporary or part-time workers to fill jobs, rather than full-time employees. We may not like these trends, but they are facts of life in the business world—facts we need to confront, not just fret over. According to one estimate, more than a quarter of American workers had "nonstandard" jobs in 2005, including temporary workers, part-time workers, free-lancers and other so-called independent contractors.[47] We're not saying that we shouldn't scrutinize these trends and think about strategies that would soften their impact on workers and communities. But we are saying that we can't expect employers to make decisions about jobs that undercut their own company's ability to compete and survive in a tough business environment. It's not going to happen.

Be Careful What You Wish For

For most Americans, low unemployment is the number one measure of whether the economy is in good shape or not,[48] and the news media are vigilant about reporting on it. Economists and business writers pore over monthly updates on how many new jobs are being created, how many jobs have been lost, and how many of those jobs are in government versus the private sector. If you want the latest details, you can find them at www.bls.gov. That's the very aptly named U.S. Bureau of Labor Statistics. You want statistics? They got 'em.

Growth, Productivity, and a Bull Market: Economic Nirvana?

But what about the other economic measures that garner so much attention? Each quarter and every year, for example, the government reports on whether the American economy is growing or not. For serious economic types, there's always buzz about productivity and where it's going. And then, of course, there's that glittering and closely watched economic indicator—the stock market. We have cable channels, websites, smartphone apps, and more offering minute-by-minute reports on how stocks are faring.

When the stars align on all three of these measures— growth, productivity, and stocks—there's generally joy on Wall Street and in economic circles. But even though they are all related to job creation and job loss, that's *not* what they measure. And sometimes, that leads to confusion.

If you aced Econ 101, you can skip this section because we're going to go back to basics for a few pages. Just to make sure that we're clear (or "crystal" as the Marines kept saying in *A Few Good Men*), we'll quickly review what these three gauges do measure and how each reflects, affects, and relates to jobs.

When the economy grows, you get more jobs, right?

One recent problem that's hitting economists upside the head is that economic growth and job creation aren't necessarily moving in tandem the way they used to. Economic growth basically measures increases in the country's total wealth, or as the experts like to say, "the value of the goods and services produced by the U.S. economy in a given time period."[49] The term you'll hear is "gross domestic product" or GDP, which is shorthand for the total economy, and economic growth means how much GDP went up or down in a given period. That includes what people, companies, and government generate.

The Bureau of Economic Analysis (a government agency) counts a whole slew of things to establish the number, including personal income and spending, corporate profits and government spending.[50] They analyze the figures to make sure changes aren't due solely to inflation. They also subtract the value of imports since, by their reasoning, these are not part of the *U.S. domestic* output.[51] Then they monitor whether the numbers go up or down.

When the economy stops growing or shrinks, it's in recession. There is a group of experts, a nonprofit called the National Bureau of Economic Research (NBER), that determines when the U. S. economy enters and leaves recessions, but the main signal NBER looks for is whether economic growth has stalled or is declining.* When growth picks up again, we're on the road to recovery.

The economic growth rate does not include measurements on job creation or unemployment—those are monitored separately by a different arm of government, the Department of Labor's Bureau of Labor Statistics. In general, a growing economy tends to produce more jobs for all the reasons we discuss throughout the book. When people spend more, businesses grow and hire more people to serve their customers. Government spending can mean that governments are hiring people or maybe they're buying goods themselves, which can also lead to jobs. And when businesses are producing more, they generally need more workers to help them do it.

But even though we often say "economic growth and jobs" in one breath as if they are one and the same, they aren't. And the problem lately is that the economy has been growing while the job situation has been lackluster or worse. We've been having a "jobless recovery."

You may have missed the official announcement, but the miserable recession of 2008 and 2009 officially ended in June 2009.[52] As we take this book to press, it's not clear whether the country is sliding back into a recession, or just having a heck of a long climb out of the earlier one. What is clear is that unemployment has remained at very dispiriting and disappointing levels, and experts like those at the CBO don't expect it to return to anything near 5 percent until 2016 at best.[53]

This has been a significant shift, and it's beginning to look like jobless recoveries might become the norm. Typically, in the recessions before 1990, unemployment started to fall four to five months after the recession was "officially" over. But in

---

* See www.nber.org.

both the 1990–91 and 2001 recessions, unemployment didn't fall for well over a year.[54] So this kind of excruciatingly slow jobs recovery is not limited to the Great Recession we've just been through.

There are several reasons why economic growth doesn't necessarily mean the jobs are coming fast and furious. One is simply lag time. Businesspeople like to make sure that good times are really here again before adding workers to their payrolls. That's just prudent from their point of view. "Historically, unemployment rates come down slowly," explains Princeton economist Alan B. Krueger. "So even with 4 percent growth, you would expect to see the unemployment rate come down maybe a percentage point a year, probably less."[55]

Once employers regain confidence, however, they still may not hire people for all the positions they had before the recession. Recessions have a way of prompting businesses to figure out how to get their work done with fewer people. They scramble; they do what they need to do to stay afloat, and they often develop new strategies for operating with fewer people. Sometimes they decide they just don't need as many employees as before. According to estimates from the Department of Commerce, U.S. business spending on equipment and software is up 26 percent since the recession, while spending on employees is up just 2 percent.[56]

A bad economy can also push marginal companies and industries over the cliff. Businesses close, and the jobs go with them.[57] The Congressional Budget Office explains that "recessions often accelerate the demise or shrinkage of less efficient and less profitable firms, especially those in declining industries and sectors."[58] There is considerable concern that the march of technology, intense foreign economic competition, and the lure of outsourcing have eliminated some U.S. jobs, even some entire occupations, forever.[59] Over the long term, this is part of the "creative destruction" we mentioned earlier. Even though the process can be heart-breaking for the people affected, it does help the country prosper over time.

So it's entirely possible for the economy to grow without creating that many jobs. Businesses only start hiring when all the existing workers are busy, there's money for invest-ment, and when the boss is convinced it's worth hiring more people.* There's a lot of debate over how fast the economy has to grow for all that to happen, although several studies say 3.5 to 4 percent.

For now, though, we'll just reiterate one key message. Economic growth is good for the country, and it would be virtually impossible to solve our jobs problem without it. But just because the economy is growing doesn't mean that all the jobs we had before are coming back. And it doesn't mean we're home free in terms of creating new jobs to take up the slack.

Rising productivity—that's a good thing, isn't it?

Productivity measures how many workers it takes to produce a company's goods or services, or, as an economist might put it, "output per worker and output per hour."[60] At its sim-plest level, it's something like this: A company that needs four workers to complete a job in a week is a lot more productive than one that has to hire eight workers to do the same thing. Being more productive cuts operating costs for the business

---

* For the economics wonks out there, this is Okun's law, after Arthur Okun, the economist who first described this in the 1960s.

and increases profits, which allows the company to grow and invest in its future. It also reduces costs for consumers and means that employers can pay their workers better wages.

If you're not the numero uno head honcho where you work, the word "productivity" sometimes seems ominous, like the boss is going to crack the whip to make you work harder. Sometimes that's the case, but big leaps in productivity generally come from technology that allows workers to do more in less time and/or better organization that saves waste and effort.

Economist Alexander Field, who wrote the "Productivity" entry for The Concise Encyclopedia of Economics (www .econlib.org/library/Enc/Productivity.html) uses changes in the U.S. auto industry in the 1920s as an example: "It took fewer and fewer hours to assemble a Model T [so] the price of automobiles fell, and the real standard of living of Americans increased."[61]

But low-tech breakthroughs can increase productivity too. "In most coffee shops," Stanford economist Paul Romer explains, "you can now use the same size lid for small, medium, and large cups of coffee." Apparently this is a fairly recent innovation (or maybe it's all the different sizes of coffee that's the innovation). Now, instead of employees hunting to find the right-sized lid or choosing the wrong one and having to do it over again, a "small change in the geometry of the cups means that a coffee shop can serve customers at lower cost."[62] Serving coffee—now more productive than ever.

The United States has enjoyed impressive productivity increases over the last 15 years or so. Productivity here rose about twice as fast as productivity in Europe, with much of the credit going to the introduction of computers and other technology.[63] We have certainly seen benefits in terms of

keeping consumer prices affordable and company profits high. (Company profits nosedived during the recession,
but they're up again now.) Sadly, American workers overall
haven't seen the rising incomes generally associated with rising productivity.[64] This change from historical patterns may
be due to international competition (something we'll examine later). Whatever the case, this pattern of rising productivity and stagnating incomes is not good news for American
workers, and it doesn't help businesses that are depending on
American consumers to buy stuff either.

Then there's the other obvious implication—when companies are more productive, they don't need to hire as many
people. Maybe you think they should keep hiring anyway, but
if they do—if they hire more people than they really need—
they'll become less competitive both here and worldwide.
Eventually, this will bring a business down.

What's more, ignoring the pluses of productivity in an
effort to hang on to jobs undercuts the economy as a whole.
Here's the way most economists explain it. Rising productivity contributes to the wealth and resources of the entire
country, and over time, it is the secret sauce that contributes
to more successful companies, higher salaries, lower consumer prices, and a rising standard of living for Americans
overall.[65] Low productivity is an economy killer. The Soviets
were really proud of their version of "full employment," which
they defined as having lots of workers in jobs even if it meant
working in wasteful, inefficient, disorganized ways. People
had jobs, but their standard of living was dismal. In time, their
whole economy imploded.

Still there is no doubt that in the short term, rising productivity can mean fewer jobs in some companies and some sectors of the economy. The question is how to tap the multiple

benefits of rising productivity and still do right by those who lose their jobs because of it.

Bull Run

The Census Bureau tracks a number of "leading economic indicators,[66] such as business income, housing starts, and data on imports and exports, and surprisingly to many, the big stock market indexes such as the Dow Jones Industrial Average, the NASDAQ, or the S & P 500 are not among the indicators measured. Maybe they think other organizations have the stock market covered. Maybe they consider it a secondary indicator, one that reflects investors' moods rather than hard facts.

On the other hand, lots of Americans do look at the stock market as one indicator of prosperity, and over the years, it has generally reflected what's happening in the economy more broadly. During the Great Depression, the Dow's value dropped by 80 percent, and it tends to tumble even in milder recession.[67] It dropped 30 percent, for example, during the relatively short 11-month recession in 1970.[68]

Similarly, higher stock prices can have a positive effect on jobs. When companies can sell their stock for higher prices, they have more money to expand and invest in new products and services that can both create new jobs and support existing ones.

A rising stock market can also create a "wealth effect." That's when people who own stocks or mutual funds see the value of their portfolios rise, and feeling so good about how wealthy they are, they go out and spend, which makes businesses more profitable and supports jobs. Funny how that works, isn't it? When the stock market is down, we tend to say, "Well, it's only on paper" (and some of us also toss unopened statements into a drawer for the duration). When stocks are going up, though, we head out shopping.

Unfortunately, the relationship between stocks and jobs is not always good, and some experts believe that corporate concern about stock prices can lead to short-term thinking and needless job layoffs. The problem is that in tough times, corporate managers often assume that a well-publicized round of layoffs will persuade investors that the company is willing to make tough choices to remain competitive and profitable.[69] Some studies suggest this isn't really so—that investors actually read layoffs as an indication that a company is in hot water and not such a good stock pick after all.[70] But that doesn't stop stressed CEOs from using this strategy to try to protect the price of their stock.

Then there's the whole bubble thing—when investors get so excited about stocks and so eager to get in on the action that they pay much more than the company is or could conceivably be worth. In the 1990s dot-com bubble, Internet companies with big ambitions, glitzy, expensive marketing and not a glimmer of profitability in sight still commanded superhuge stock prices—until one day they didn't. The bubble burst, stock values fell, the companies went out of business, and the jobs of their employees went down the tubes with them. In that case, soaring stock prices turned out to be very bad for jobs and people's 401(k)s as well.

But like the other indicators we've discussed here, the value of the stock market is not an accurate measure of how well the country is doing on jobs. Essentially, the Dow and other market indexes only reflect how much people are willing to pay for shares in the publicly traded companies they track.

Close, but No Cigar

Our main point is that it's a mistake to assume a growing economy, better productivity, and a rising Dow will take care of our jobs problems—close, but no cigar.* These are all indications that some good things are happening in the economy, but they don't necessarily mean we're reclaiming or creating enough good jobs to get ourselves out of the hole we've fallen into.

---

* We were disappointed to learn that "Close, but no cigar" isn't a quote from Groucho Marx. Apparently it's from the custom of giving cigars as prizes at state fairs some years ago (www.phrases.org.uk/meanings/close-but-no-cigar.html). One online source we checked admits "there's no definitive evidence to prove" the fairground theory. In a just world, it really would be Groucho.

Before They Hit the Big Time

Some of the companies that have contributed the most American jobs started out with very humble beginnings. Their founders made billions because they saw a business opportunity and decided to go for it. And as their companies grew and expanded, they created more and more jobs. Here are some of corporate America's most astonishing success stories.

WALMART

The small start: Sam Walton worked in a JCPenney after leaving the military and opened Walton's Five & Dime in Bentonville, Arkansas, in the early 1950s. But he didn't stop there. He opened the first Walmart in Rogers, Arkansas, in 1962.[71]

Photo: Susan Wolfe.

The big time: Now "the world's largest retailer," Walmart has more than four thousand stores nationwide and made more than $400 billion in sales in 2010.[72] Not only is Walmart the biggest employer in the United States with 1.4 million "associates" nationwide, it's the biggest employer in Mexico and one of the biggest in Canada too.[73] Critics claim that Walmart underpays its workers and that the company often puts smaller retailers that cannot match its low prices out of business.[74] According to Walmart, its average full-time wage is $11.75 per hour. More than 800,000 of its employees are women, and the Walmart greeter cliché isn't that far off the mark—more than 400,000 of Walmart's U.S. employees are over fifty. The company employs some 2.1 million people worldwide.[75]

MARTHA STEWART OMNIMEDIA

Photo: Susan Wolfe.

The small start: She was a model and a stockbroker before she became the queen of elegant domesticity, but Martha Stewart set the stage for her media empire when she opened a catering business and in 1982 published the first of many books of recipes and homemaking advice.[76]

The big time: Amazingly, Martha Stewart's company, which publishes books, produces TV programs, and merchandises all things Martha, generated nearly $245 million in revenues in 2009 with only 620 employees.[77] But that's not the whole story. At Martha Stewart Crafts online, for instance, you can order "medium pink pom-poms" (they seem to be used to decorate the dining table) for $12.99, along with a 24-pack of glitter glue for $29.99. The folks who sell and ship those products hire people too, and we're willing to bet that the pink pom-pom and glitter glue businesses are a lot more profitable now than they were before Martha came on the scene. In fact, we're not sure how many people knew they needed pink pom-poms before Martha came along, but now they do, and it all creates jobs.

FRITO-LAY

Photo: Susan Wolfe.

The small start: Sometimes it takes two to tango, and this snack powerhouse is the result of two entrepreneurial forces. In 1932, Herman Lay was delivering snacks in Tennessee and decided he could do better by buying the company and running things himself. Thousands of miles away, C. E. Doolin munched on some chips in a San Antonio café and decided to buy the recipe for Fritos.[78] The two snack visionaries joined forces in 1961, thus ensuring that American sports fans never have to go hungry while cheering their teams onward.

The big time: Frito-Lay has been called the "undisputed chip champ of North America," and the company employs some forty-eight thousand people making Doritos, Tostitos, Ruffles, and Cracker Jack, among other munchies.[79] Frito-Lay is now a subsidiary of PepsiCo, providing a quarter of its sales.

AMAZON.COM

The small start: Jeff Bezos was working on Wall Street in the 1980s and 1990s and watching the growing influence of the Internet when he decided that there was money to be earned selling books online. He set up the mail order business in his garage with three computers purchased at Home Depot.[80]

The big time: Amazon.com was an immediate hit. Within a month of its launch, it had found customers in all fifty states and abroad as well.[81] Now, of course, Amazon sells all sorts of products, from books to clothing to food. In 2010, the company brought in more than $34 billion in revenue.[82] It has about thirty-three thousand employees worldwide, and U.S. fulfillment centers—where employees package and ship the merchandise—in eighteen states.[83]

Photo: Susan Wolfe.

INQUIRING MINDS WANT TO KNOW
Sorting Out the Big Debate over Jobs

If you spend any time at all listening to what different experts and leaders think should be done to improve the U.S. jobs picture, you might find this image popping into your mind: a crying baby. All the adults are scurrying around giving their advice on what to do to make the baby happier. The baby's too hot; take off his sweater. No, he's too cold; he needs a sweater. He just needs changing. Maybe he's teething or has an earache. Maybe it's colic. Pick him up and walk around for a while. No, just let him cry a little—he'll be asleep all by himself in few minutes. Maybe it's serious; maybe we should call the doctor.

The suggestions on what the country should do to create more jobs here in the United States and generate more good ones are at least as numerous and just about as contradictory. In the following chapters, we'll work our way through the main ones, concentrating on those that are likely to be in play, both in the upcoming elections and for years afterward. Most of these ideas seem plausible, and most Americans, including your authors, have wondered about exactly how these ideas work and which of them is most likely to get the American jobs machine humming again.

So here's our question in the chapters to come: Would doing X hurt or help when it comes to jobs?

The Public's Starting Point on Jobs

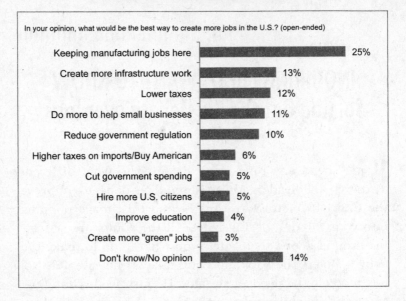

In your opinion, what would be the best way to create more jobs in the U.S.? (open-ended)

Keeping manufacturing jobs here	25%
Create more infrastructure work	13%
Lower taxes	12%
Do more to help small businesses	11%
Reduce government regulation	10%
Higher taxes on imports/Buy American	6%
Cut government spending	5%
Hire more U.S. citizens	5%
Improve education	4%
Create more "green" jobs	3%
Don't know/No opinion	14%

SOURCE: Gallup survey, March 25-27, 2011.

How do your views of jobs solutions compare to what most Americans are thinking? And after you've gone over the coming chapters, have you changed your mind about any of these ideas? We certainly reconsidered our own views while we were working on the book—you're welcome to join us. We consider it the smart thing to do.

CHAPTER 5

WOULD BALANCING THE BUDGET HELP CREATE JOBS?

The time to hesitate is through.

—*Jim Morrison*[1]

Alfred Hitchcock used to say that he learned about suspense as a schoolboy. His school, like many others at the time, used corporal punishment—the boys were smacked on the hand with a ruler whenever they misbehaved.

But Hitchcock's school added a twist—the student himself chose the time when the punishment would be given: before class, after class, during recess.[2] For Hitchcock, knowing what was going to happen, but being able to postpone the dreaded moment until later, made the experience even more harrowing.

THIS IS GOING TO HURT

That's how a lot of us feel about the country's budget problems—the routine deficits and gargantuan debt the United States has amassed. We know we have to solve this problem, and we know solving it is going to hurt, so we seem to keep postponing the most painful choices. The Congress and President Obama have taken some initial steps, but with federal debt now speeding past $15 trillion, this issue will be with us for a long, long time.

There is a group of economists and commentators who argue that in the wake of the Great Recession, deficits are just what the doctor ordered—or at least, what a doctor of economics would order. With the private sector still sluggish, higher government spending and lower government taxes are the best way of getting the economy moving, they argue. But there are also those who believe the best strategy is to reduce our deficit spending as quickly as possible, which of course means cutting government programs and raising revenues. So that brings up the question of which should take precedence: attacking the deficit and the debt, or doing whatever it takes to create and hold on to jobs even if it means delaying action on the budget or maybe even making our budget problems worse for a while. Welcome to the stimulus-versus-austerity debate.

The nonpartisan Congressional Budget Office has been clear about the dilemma. There are ideas on taxes (lowering them) and spending (extending unemployment benefits, investing in infrastructure projects, and the like) that might help on the jobs front in the near term. "But there would be a price to pay," the CBO says. "Those same fiscal policy options would increase federal debt, which is already larger relative to the size of the economy than it has been in more than 50 years—and is headed higher."[3]

Just to make things more confusing, economists generally argue that for jobs and the broader economy, a government deficit can be good or bad, depending on the overall circumstances. So where are we now? Let's look at what we know for sure.*

GETTING YOUR BEARINGS: A FEW FACTS TO CONSIDER

There's not much dispute about basics. The Congressional Budget Office, the Government Accountability Office (GAO), the Office of Management and Budget (OMB), and various commissions and bipartisan groups have been sounding the alarm on the country's budget problems throughout the George W. Bush and Obama administrations.

- The United States has spent more than it has collected in taxes for thirty-one out of the last thirty-five years, but the Great Recession put deficit spending into overdrive. In 2010, we spent $1.3 trillion more than we collected in taxes. Current estimates show that between 2012 and 2021, the country's deficits will total more than $5 trillion.[4]
- Even more disturbing, we've got deficits built into the future as well. As the baby boomers get older and health care costs keep going up, the government will have enormous trouble keeping up with the bills for Social Security, Medicaid, and, more than anything else, Medicare. According to the Congressional Budget Office, "the aging of the population and the rising cost of health care will cause spending on the major mandatory health care

---

* Just for background: The government runs a *deficit* when it spends more money than it takes in by taxes in a given year. When that happens, the government makes up the difference by borrowing. The total amount the government has borrowed is the *national debt*. Whenever you buy a Treasury bond, you're lending money to the government.

programs and Social Security to grow from roughly 10 percent of GDP today to about 16 percent of GDP 25 years from now if current laws are not changed."[5] The CBO pointedly notes that 16 percent of GDP is close to what we now spend on *all* government programs, including defense.[6] Number crunchers sometimes call what the country owes down the line for these programs "unfunded liabilities," but there are human beings behind the figures—Americans who have paid Social Security and Medicare taxes throughout their working lives and now or will soon need the help these programs provide.

Government's Financial Burdens Soar after 2020

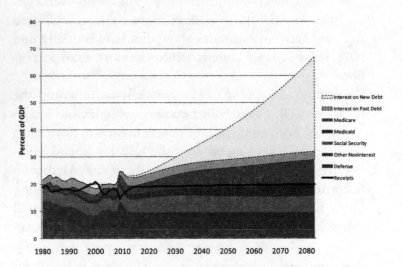

SOURCE: U.S. Treasury, "A Citizen's Guide to the Financial Report of the United States Government," 2009.

This is what the government's own auditors mean when they say the federal budget is "unsustainable." After 2020, the government will have increasing trouble covering its bills, as it faces rising costs on health care, Social Security, and just keeping up with the interest on its debt.

- Our budget problems are so enormous they can't be fixed by cutting what people normally think of as waste, fraud, or "big government." Eliminating the Departments of Education, Agriculture, Energy, Transportation, Justice, and Housing in 2010 would have saved about $300 billion. But the deficit that year was $1.3 trillion. Even if we axed these departments entirely, not just cut them back, we still would have been $1 trillion in the red. What's more, many Americans believe that these departments provide important and needed public services.

- The recession put us much deeper in the hole. Deficits usually go up during bad economic times, because unemployed people and bankrupt businesses don't pay as much in taxes. Plus, the government spent more money on unemployment benefits, help for state and local governments, and trying to stimulate the economy back to health. Unfortunately, returning to prosperity won't be enough to solve our budget problems. The CBO, GAO, and OMB all agree that the costs for Medicare, Social Security, and interest on the debt will grow faster than the economy, even if we hit boom times again. Alan Simpson, the cochair of the White House budget commission, tends to say things that aggravate people, but the man is spot-on here: "Hell, we could have double[-digit] growth for 30 years," Simpson said in 2010, "and never grow our way out of this."[7]

- The United States issues Treasury notes to cover what we borrow, and about a third of these are now held by the governments of foreign countries, including China, Japan, the United Kingdom, and Russia. Like any other borrower, the country pays interest on the borrowed money. Based on CBO estimates, interest alone could cost the country more than $5 trillion between 2011 and 2020.[8]

We published a "guided tour" book similar to this one on the budget issue in 2011,[9] and in it we relied on the sage economic advice we used as our opening quote for this chapter—"the time to hesitate is through," as Jim Morrison and the Doors so deftly put it. But we need to act on jobs too. The vast majority of Americans want progress on both jobs and the deficit: 80 percent say it's a "very important" priority for Congress to act on jobs, while 70 percent say the same about reducing the deficit.[10]

Sad to say, economists and political leaders are bitterly divided on how to balance these two goals. Some argue that the budget issue is the top priority and that addressing it will be better for jobs in the long run. Others say that unless we put jobs first, the United States risks sliding into an economic wasteland that would impoverish millions. We wish there were a simple, clear answer on this, but there's not.

SO HOW COULD CLOSING THE BUDGET GAP HELP ON JOBS?

Experts who believe that the budget issue is the top priority worry that *not* acting on it is so risky that we simply have no other choice. Basically, they say that unless we start controlling the deficits, we'll see even worse levels of economic pain and unemployment in the future.

The most alarming possibility is a debt crisis like those that battered Greece and Ireland in 2010 and Argentina in 2001. In a debt crisis, international investors holding a country's bonds start to worry that said country is in over its head, and they begin moving their money elsewhere. The country still needs to borrow to keep everything afloat, but it can't find new investors to keep the money flowing.

So far, we haven't had much trouble finding people to lend us money. Up to now, the United Sates has been seen as one of the safest investments in the world—which is remarkable, given how hard our leaders in Washington have tried to throw that advantage away lately. The completely artificial crisis in 2011 over raising the national debt ceiling (the total amount the government is allowed to borrow) caused Standard & Poor's to downgrade the U.S. government's credit rating from the top-of-the-scale AAA rating. And although Moody's, S&P's rival bond rating agency, didn't downgrade the government, they gave it a "negative outlook," meaning that we're okay now but headed for trouble.

The bond rating firms have come in for bitter criticism because of their failure to predict or prevent the financial crisis in 2008. They deserve it. But the most interesting perspective is why the two agencies differ on this. Moody's argues that the enormous size of the U.S. economy, plus the dollar's role as the keystone of international finance, means that the U.S. government is good for the money it borrows.[11] S&P, on the other hand, says the debt ceiling debate shows our political system can't handle this problem. "The political brinksmanship of recent months highlights what we see as America's governance and policymaking becoming less stable, less effective, and less predictable than what we previously believed," S&P said.[12] In other words, the U.S. government could pay its bills, but maybe it won't. If the financial world as a whole moved to S&P's view, that would be as bad for the U.S. economy as actually running out of cash.

What would happen if a debt crisis hit the world's largest economy, as opposed to smaller ones such as those of Greece and Ireland? It's uncharted territory. The possibilities include a plummeting stock market, a shell-shocked world economy, skyrocketing interest rates on everything from small business loans to mortgages, and a mad scramble by the U.S. government to hike taxes and slash spending in a frantic effort to reassure international investors. And it could arise out of the blue, just like the mortgage and banking crises that nearly tumbled the economy in 2008. Everything was going along fine until suddenly it wasn't.

But there are other, less dramatic ways a rising national debt could hurt the economy. One possibility—and one that is arguably more likely than a full-blown debt crisis— would be that all the government borrowing would start sucking up money that would otherwise be going into private investment. It's basically supply and demand. There's only so much money available for investment, and if too much of it is used for government debt, there's less for private firms—which results in driving up interest rates for everybody. Higher interest rates mean that it's more costly for entrepreneurs to obtain loans to start and expand businesses (which add jobs) or for consumers to get loans for large purchases such as houses and cars (which also support jobs). And of course, raising taxes and cutting spending to deal with the debt has implications for the overall economy too.[13]

There's one critical point here, however. Nobody knows exactly when these bad effects might happen, or how bad they might be. Controlling the budget is about reducing the risks.

AUSTERITY OR STIMULUS? THAT IS THE QUESTION

The budget-versus-jobs debate is often couched as a battle between austerity and stimulus. Austerity is just what it sounds like—cutting spending and raising taxes to try to get the budget in line—and no one is suggesting that it's fun.

Many Americans tend to equate the stimulus with the big bank and auto industry bailouts in 2008 and 2009, but that was actually something different—the so-called TARP legislation (*TARP* stands for Troubled Assets Relief Program—and *troubled* is probably one of the great economic understatements of the decade). The true meaning of stimulus is government spending that's intended to spur economic and job growth. The granddaddy of stimulus programs was Franklin Roosevelt's New Deal during the Great Depression, which tried to put people to work using massive public works projects, farm subsidies, and relief programs such as the Civilian Conservation Corps. As we mentioned in Chapter 4, President Obama's $787 billion stimulus package in 2009 included tax cuts, money sent to the states and cities to help them weather the poor economy, help for the unemployed, and money to jump-start a variety of projects thought to provide jobs. The bipartisan tax deal reached at the end of 2010 ended up looking a lot like a stimulus program too, not only extending the so-called Bush tax cuts for two years but also throwing in a one-year payroll tax holiday.*

Surveys show that the public is skeptical about how much the stimulus accomplished, but the best nonpartisan estimates from the CBO say the first round of stimulus reduced the unemployment rate by somewhere between 0.8 and 2.0 percentage points (this at a time when overall unemployment was hovering near 10 percent).[14]

The firefight over austerity versus stimulus has ignited some knock-down drag-outs among big economic icons. Former Federal Reserve chairman Alan Greenspan is promoting the austerity track, as is Harvard economist Ken Rogoff.[15] The International Monetary Fund is an austerity

---

* If you're looking for the nitty-gritty on the bailouts, stimulus, health care bill, and defense spending, there's more in our earlier book. The details are at www.wheredoesthemoneygo.com.

proponent, and in the wake of the European debt crisis of 2010, some countries (the United Kingdom is probably the best example) seem to be taking that advice to heart. The British are phasing in major spending cuts—the largest since World War II—reducing budgets for everything from defense to policing to higher education and the BBC. The country also raised its value-added tax (a kind of sales tax) from 17.5 to 20 percent.[16]

In the United States, most austerity proposals focus on cutting spending rather than raising taxes, and even those that feature both rely more on spending cuts than tax hikes, at least so far. American experts have typically been wary of raising taxes when the economy is sluggish (witness the number of prominent economists who backed the agreement to extend the Bush tax cuts for two years at the close of 2010).[17] Elected officials hardly ever like the idea of raising taxes.

For the austerity camp, the country's budget problems are like a sword hanging over the economy's head. They believe the risks associated with carrying so much government debt make investors and businesspeople cautious about putting their money into new enterprises or expanding old ones.[18] Reducing the deficit, they argue, would give these key groups confidence that the country is moving in the right direction and protect us against a disastrous economic future.

Some analysts raise more fundamental objections. They doubt whether government can or should step in to promote job creation, even if we could afford it. For example, Dan Mitchell at the Cato Institute writes that "there is virtually no support for the notion that government spending creates jobs. Indeed, the more relevant consideration is the degree to which bigger government destroys jobs."[19]

For Mitchell and others, cutting the size of government—

not just addressing the debt—should be our real goal. They believe that the United States would be far better off with a much smaller government. Their objective is not just to stem the red ink. It's to do it by dramatically reducing the size and scope of the federal government.

Not surprisingly, small-government advocates believe that leaving money in private hands is far better for the economy and job creation. The Club for Growth, for one, contends that there is "a huge economic cost" when the "federal government extracts nearly $3 trillion in federal taxes from families and businesses each year."[20]

The theme of government and taxes and how different ideas about them play out in terms of jobs is one we'll continue to explore throughout the book. For now, let's just leave it at this: some experts want to make a beeline to tackle the country's budget problems to avoid an economic catastrophe in the future, and some portion of those believe the way to do so is by making government substantially smaller.

CAN THIS IDEA SAVE THE DAY?

We wouldn't have spent so much time writing about the austerity-versus-stimulus debate if there weren't another side to the story, and a very important one. There are plenty of reputable economists and analysts who believe that now is not the time for major spending cuts—not when the economy is performing so pitifully and when unemployment is still high.[21]

For many observers, rushing to cut the deficit when millions of Americans can't find jobs is focusing on exactly the wrong problem. Columnist Joe Klein, who has applauded some ideas for attacking the country's budget problems

and criticized others, still believes the debate is off course.*
"There is a larger problem: Why are we spending so much
time and effort bloviating about long-term deficits and so
little trying to untangle the immediate economic mess that
we're in?"[22] Katrina vanden Heuvel, the editor and publisher
of the *Nation*, made the same point when she critiqued
President Obama's approach to the budget debate, comparing it to that of the Republicans: "Why is the debate we are
having not about whether to cut, but how much to cut? Why
isn't it about the urgency of joblessness instead of the perils
of deficits?" Vanden Heuvel considers the president's plan
"an austerity-lite policy, one that all but guarantees mass
unemployment as the new normal."[23] When a coalition of
progressive groups created their deficit-reduction plan in
2010, one key proposal was that the government shouldn't
even worry about cutting spending until unemployment
dropped to 6 percent or below for six months.[24]

Others are even more skeptical about proposals for controlling the budget. Economist Robert Kuttner worries that
the various debt reduction plans are thinly veiled attempts
to slash government and balance the budget on the backs
of the middle class and the poor. When the *New York Times*
published an exercise inviting readers to try their hand at
reducing the deficit, Kuttner saw the whole enterprise as
"insidious [because] they take the premise of the deficit
hawks for granted—that the projected deficit rather than
the prolonged slump is the top economic challenge."[25]

Most of those who recoil against making the budget the
top priority believe that more stimulus spending is essen-

---

* This column was written in response to the so-called Simpson-
Bowles proposals in December 2010. You can review their ideas at
www.fiscalcommission.gov/sites/fiscalcommission.gov/files/docu
ments/TheMomentofTruth12_1_2010.pdf.

tial to getting the country out of its economic doldrums. If we don't do this, they argue, fixing the budget will become even harder. Helping small businesses and investing more in job training and education are top priorities. Extending unemployment benefits is crucial as well. Not only does this help individuals, it helps local communities and businesses too. (People who don't have money don't spend it, and that means fewer customers.)

Then there's the additional negative impact on jobs: cutting government spending may sound totally benign, but it means some people will definitely lose their jobs. It would be foolish to pretend otherwise. Part of the Obama stimulus package in 2009 went to state and local governments to help them keep teachers, police officers, firefighters, and other government workers on the payroll. When that stimulus money ran out, the layoffs started.

And cutting federal spending can lead to substantial job losses. In early 2011, when some in Congress proposed immediate, extensive spending cuts and there was talk of a government shutdown, Mark Zandi of Moody's Analytics (who advised Senator John McCain during the 2008 presidential campaign) warned that abrupt actions such as this could cost the economy some 700,000 jobs through 2012. "Significant government spending restraint is vital," Zandi wrote, "but given the economy's halting recovery, it would be counterproductive for that restraint to begin until the U.S. is creating enough jobs to lower the unemployment rate. Shutting the government for long would put the recovery at risk, not only because of the disruption to public services but also because of the potential damage to consumer, business and investor confidence."[26]

Nobel Prize–winning economist Paul Krugman has written repeatedly and passionately on why government needs to spend more now to stimulate the economy and create jobs. In fact, Krugman advocated a much larger stimulus package back in 2009. It would have made the deficit that year much larger, but Krugman believes that the economic recovery would have been far stronger if we hadn't been so penny-wise and pound-foolish.

Of course, Krugman doesn't deny the importance of addressing the budget mess later. "Looking ahead, we're going to have to find a way to run smaller, not larger, deficits," he wrote in 2010.[27] But he also says that we simply have to "combine actions that create jobs now with other actions that will reduce deficits later."[28]

CAN WE HAVE A LITTLE OF BOTH, PLEASE?

The problem is how to put together a package that addresses both challenges squarely. Economists—and citizens too—are worlds apart on the specifics, but there is a fairly broad agreement that we'll need some sort of mix-and-match solution. Alice Rivlin, a highly regarded budget expert and member of the White House fiscal commission,* believes that blending the two priorities is essential: "Any serious plan to cut the massive debt build-up requires reining in future federal spending and raising revenues. Both can be done in ways that give us a more productive economy and keep the recovery going."[29] What bothers Rivlin is the "buzz saw of exaggeration" that seems to accompany any specific idea put on the table. "The left screams that reducing federal spending growth will hurt the poor, the

---

* Alice Rivlin also served as a board member of Public Agenda and an expert reviewer for our book on the budget issue, *Where Does the Money Go? Your Guided Tour to the Federal Budget Crisis* (New York: HarperCollins, 2011).

elderly, and working families. The right shrieks that tax increases will kill investment and enterprise."[30]

She's not the only one. A number of blue-ribbon panels that have examined the nation's fiscal problems, such as the Committee on the Fiscal Future convened by the National Research Council, the Peterson-Pew Commission, and the Bipartisan Policy Center, have all concluded that we can and should combine short-term spending to help the economy with long-term deficit reduction. Even organizations that are completely committed to controlling the debt, such as the Concord Coalition and the Committee for a Responsible Federal Budget, say the same.

In the spirit of full disclosure, we want you to know that we're in the "we have to do both" camp ourselves. And like Rivlin, we're worried that politics is getting in the way of making sensible decisions. "There's absolutely no reason," we wrote in 2011, "we can't do two things at once, unless of course, our political culture is so partisan and gridlocked that it effectively can't walk and chew gum at the same time."[31]

At the top of this chapter, we posed a question about whether balancing the budget, or at least reducing the deficit, will help or hurt on jobs. And the answer, from the economists and experts we've talked to, is yes on both scores. If we allow our national debt to rise to unsustainable levels—and that could happen in as little as ten to twenty years—that debt burden will drag down the overall economy, costing jobs.

But arguably, if we're too aggressive in cutting the federal budget now, that could cost jobs too. For one thing, it'll cost the jobs of federal workers who get laid off and may or may not have jobs waiting for them in the private sector. And for another, government spending is helping to prop up the private sector in the short term.

We're going to be living with this austerity-versus-stimulus debate for a long time, because both our budget problems and our jobs problems are long-term, persistent challenges. We need to chart a course that gets us past both risks, because if we solve only one of them, the other one will get us for sure. It's possible to find a solution to both. But we have to be a lot smarter and a lot more flexible than our political system has been for the last few years.

||||||||||||||||||||||||||||||||||||||||||||||||||||||||||||||||||||||||||||

CHAPTER 6

WOULD CUTTING TAXES HELP CREATE JOBS?

I'm proud to pay taxes in the United States;
the only thing is, I could be just as proud for half the
money.

—*Arthur Godfrey*

From a politician's perspective, cutting taxes to help get the country out of its jobs doldrums is the twofer of all time. Cutting taxes is bound to please most voters, and you can say you're tackling the public's biggest worry—the lack of jobs. When President Obama and the Republicans made their bipartisan deal on extending the Bush tax cuts (plus a temporary cut in the payroll tax) they both argued for tax cuts as one means to getting the economy moving and employers hiring again. But from an economist's perspective, things are a little more complex.

First off, how exactly does cutting taxes create jobs, and which taxes should be on the chopping block? To spur hiring, do you cut income taxes, business taxes, taxes on the middle class, taxes on savers and investors, or taxes for everyone? And given the country's humongous debts and the gushing red ink in the federal budget, is it possible that cutting taxes might do more harm than good?

GETTING YOUR BEARINGS: A FEW FACTS TO CONSIDER

Given all the hoopla about taxes—and the number of times that political candidates say that cutting them is one of their goals—it's probably advisable for all of us to take a deep breath and get a few facts under our belts before moving on. Here's a mini refresher course in where we are on taxes. We'll just speed right through with the help of a few charts.

- Most of the taxes the federal government collects come from two sources. In 2010, income taxes made up 41.5 percent of the total, and payroll taxes 40 percent. Corporate taxes accounted for 8.9 percent, and excise taxes on gas, booze, and cigarettes another 3.1 percent.* Some other miscellaneous taxes and fees make up the rest. Depending on where you live, you probably pay additional sales and/or real estate taxes to state and local governments, as well, but most of the national political debate focuses on federal taxes.

---

* The exact breakdown can change depending on the economy and how tax laws are changed over time. The percentages have been in this ballpark for at least the last decade, but you can see for yourself: Office of Management and Budget, "Historical Tables," Table 2-2: "Percentage Composition of Receipts by Source, 1934–2016," www .whitehouse.gov/sites/default/files/omb/budget/fy2012/assets/ hist02z2.xls.

Sources of Federal Revenue, 2010

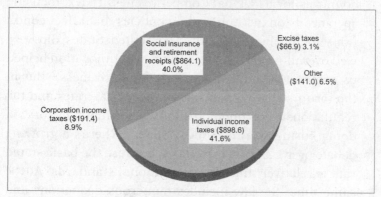

Social insurance
and retirement
receipts ($864.1)
40.0%

Excise taxes
($66.9) 3.1%

Other
($141.0) 6.5%

Corporation income
taxes ($191.4)
8.9%

Individual income
taxes ($898.6)
41.6%

SOURCE: *Budget of the United States Government, Fiscal Year 2012.*

Personal income taxes and social insurance receipts (mostly payroll taxes for Social Security and Medicare) are the federal government's main source of income.

- It may not seem like it come April 15, but federal income taxes are actually low compared with where they've been in the past. Both the Reagan and George W. Bush administrations cut income taxes substantially, and our income taxes are generally lower than what people pay in Western Europe. In fact, according to the nonpartisan Tax Policy Center, the overall tax burden in the United States (everything we all pay in all sorts of taxes, compared to the size of our economy) is one of the lowest among developed countries—just 27.3 percent of our gross domestic product, compared to more than 40 percent in countries such as France, Sweden, and Denmark.[1] (See the chart on page 106 for the details.) On the other hand, some studies looking at international economic competitiveness suggest that the United States may have some problems with its tax policies for business. The World Economic Forum rated more than 130 countries for its Global Competitiveness Index, focusing specifically on institutions and policies that affect conditions for business growth. The United States did very well overall—open markets and lots of innovation helped us come in at number four in competitiveness.[2] But in the forum's survey of business leaders, tax rates and tax regulations were among the top "problematic factors" in doing business in the United States.[3] There's a growing debate over U.S. corporate taxes because the basic stated rate is relatively high by international standards. At the same time, the law includes numerous deductions and loopholes, so some large companies actually end up paying very little.[4] We take up the idea of reforming corporate rate taxes in Chapter 16.

International Tax Burdens

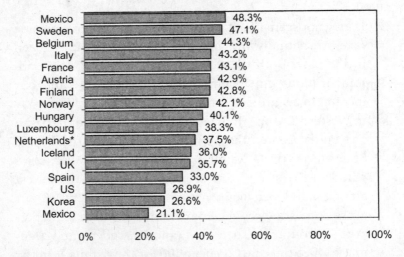

Country	Percentage
Mexico	48.3%
Sweden	47.1%
Belgium	44.3%
Italy	43.2%
France	43.1%
Austria	42.9%
Finland	42.8%
Norway	42.1%
Hungary	40.1%
Luxembourg	38.3%
Netherlands*	37.5%
Iceland	36.0%
UK	35.7%
Spain	33.0%
US	26.9%
Korea	26.6%
Mexico	21.1%

SOURCE: Organisation for Economic Co-operation and Development.

When you look at the tax burden (including state and local taxes) as a percentage of the total economy, people in other countries pay a much bigger share to their governments.

- The non-partisan Tax Policy Center estimates that about half of Americans (47 percent in 2009) don't pay income taxes, and no—despite everything you've heard about Wesley Snipes and Willie Nelson—we don't have that many cheats and scoundrels.[5] It's mainly because millions of Americans don't earn enough to pay income taxes. That doesn't mean you're off the hook entirely. Most lower-income Americans still have to pay taxes for Social Security and Medicare (those ever-popular payroll taxes), and nearly all pay sales taxes where they live.

SO HOW WOULD CUTTING TAXES HELP ON JOBS?

Proposals to cut taxes to spur job creation come in a lot of different flavors, but there are basically two varieties. Some proposals rest on the belief that cutting taxes on what people earn spurs job creation, since consumers have more money to spend on products and services. Others center on cutting taxes (or giving tax breaks) to businesses to make it easier and cheaper for them to hire more people.

Putting More Money in Consumers' Pockets

Shopaholics can pat themselves on the back in at least one regard—buying things is a big part of what makes the U.S. economy tick. Consumer spending is estimated to be about 80 percent of spending in the private sector and two-thirds of spending overall,[6] so when shoppers put themselves on a budget, it's not good for jobs. Given this, one way to preserve and create jobs is to cut taxes so that consumer spending rises.

When people keep more of their pay, they are more likely to drop by Target on the way home or head out to Olive Garden on Friday night. These businesses then need more cashiers and salespeople; the companies that produce whatever is being sold (be it bathroom towels or the ingredients for the Never-Ending Pasta Bowl) need more workers. Stuff has to be shipped around the country to get to consumers, so FedEx and UPS need more truck drivers and workers. If the going really gets good, companies open new branches (is there a new Target or Olive Garden near you?), for which they hire architects and builders. Then they hire more workers to take care of customers who come into the new locations. (Yes, we too have noticed that annoying trend of having customers take care of themselves—scanning their own purchases and packing their own bags. If the Olive Garden sends you into the kitchen to shred the Parmesan, we'll know it's gone way too far.)

Since the idea is for people to keep more money so they can spend more, this strategy emphasizes cutting the taxes paid by individuals—typically income and payroll taxes. Conceivably you could slash taxes on gas, cigarettes, and alcohol as well. But as a nation, we've already made a public health decision to discourage smoking and drinking as much as possible. And encouraging people to drive more won't do much for global warming or our dependence on foreign oil (plus there's the fact that gas taxes fund most of the road and bridge maintenance). So, politically speaking, the most visible ideas are to cut income and payroll taxes. And keep in mind that these two categories make up roughly 80 percent of the taxes Uncle Sam collects. They really are where the action is.

There are a couple of complications, of course. One is whether and how quickly people will actually spend the money. If you're counting on the tax cut to ramp up consumer spending, people need to spend. This strategy is not so effective if people instead use the money to pay off their debts, stash it away in some long-term investment, or just stick it in the bank because they're worried about the future. Those might be good and responsible things to do from the individual's point of view—and there are many economists who say that we're going to have to pay off debt and encourage more saving and investing to have a strong economy in the long run—but it doesn't help create jobs quickly.

There's also a huge amount of frothing at the mouth from politicians over whether the chief goal should be to keep taxes low for lower-income and middle-income people or to keep taxes low for high-wage earners as well. Depending on your own tax bracket, this controversy could hit close to home.

For Steven Pearlstein, who writes for the *Washington Post*, the "economics here are pretty straightforward. [For the middle class in a fragile economy] nearly every dollar taxed is nearly a dollar not spent buying goods and services."[7] But nobody spends like the top earners. According to Moody's Economy.com, about 45 percent of consumer spending comes from people who have incomes in the top 25 percent.[8] Many economists believe that until the U.S. economy is a lot stronger than it has been for the past few years, we really need everyone, rich, poor, and in between, out there spending money.

What's more, some analysts argue that taxing high earners undercuts their incentive to invest in new ventures that could produce even more jobs in the long run.[9] Economists who want to make this point typically veer into nearly incomprehensible economic verbiage, but it's easy to make this concrete. Think Steven Spielberg. He's fabulously wealthy, and he doesn't need to work to put food on the table. But he keeps making movies, and he creates thousands of jobs by doing so. Steve Jobs was another example. He achieved some of his most astonishing breakthroughs after he was already a rich and famous man. The problem is that if we make taxes on the wealthy too high, they might decide that it's just not worth it anymore. Why knock yourself out when you have to hand over to the government a huge portion (and "huge" is obviously in the eye of the beholder) of what you make?

Or why stay here and pay the tax man when you can pick up and move to the Cayman Islands or somewhere else where people can keep more of what they earn? It's not an unknown phenomenon among the wealthy and successful. Rod Stewart and the Rolling Stones left the United Kingdom to avoid paying high taxes there.[10] For Stewart at least (he mainly lives in California) even U.S. taxes apparently looked good by comparison.

Is Cutting Taxes to Encourage Spending Really Such a Good Idea?

But let's say you could craft a tax cut plan that gets as many people out there spending and working as possible. Even so, not everyone agrees that this is the way to go. Alan Greenspan, for one, says the country needs to raise taxes to reduce the country's deficit and debt, and he's not alone. Bruce Bartlett and David Stockman, both economic advisers to President Reagan, make exactly the same point.[11] Others say the money would create more jobs if it were invested in building bridges and airports and/or in moving the country much more quickly into wind and solar power and other forms of green energy.[12] (That's a strategy we take up in Chapter 10.)

Another complication is that the idea of cutting taxes to encourage consumer spending runs right up against another key financial goal—getting American families to reduce their debt and start saving again. When the Great Recession hit, consumer debt was at record levels and saving at near record lows,[13] so there is a genuine tension between getting Americans to spend more to pump up the economy and create jobs and having them get their own personal finances under control. For some, counting on American consumers to go on a spending spree is like asking marathon runners to keep on running after the 26.2-mile mark. In the wake of the recession, many Americans have been using any extra money they have to pay off debt, not to buy more.[14] The problem here is that many Americans can't do both—spend more to pull the economy out of its doldrums and pay down debt and save more to protect their own futures.

Making It Cheaper to Add Jobs

But individual Americans aren't the only ones who pay taxes, and a number of economists believe that the better and more direct approach to creating jobs is to cut the taxes that make it more expensive for businesses to hire. Remember, when employers hire new workers, they're not just paying salaries; in most cases they're also paying for benefits such as health insurance, and various taxes as well, such as the employer's share of Social Security, Medicare, and unemployment insurance. The total cost of a new employee is usually 25 percent to 40 percent more than the actual salary.[15] Leaders who emphasize cutting business taxes say this is a more direct way to spur hiring because it makes it cheaper to bring more people on board.

The nonpartisan Congressional Budget Office calculated how many jobs would be created by various kinds of tax cuts, both for individuals and for businesses. Their study didn't look at every possibility, but it did compare some of the main ideas now being considered. The wonks will have their way with the English language, so, predictably, the CBO number crunchers talk about "years of full-time equivalent employment per million dollars of total budgetary cost."[16] It doesn't roll trippingly off the tongue, but the idea is vital, and it means pretty much what it says—the number of full-time jobs that would be created per million dollars of taxes that don't come into the U.S. Treasury. Unless you're Bernie Madoff, money can't be in two places at one time, so cutting taxes means there will be less money coming into the U.S. Treasury.* It's a crucial point given how serious the country's overall budget problems are.

The CBO results are eye-popping. Based on their calculations, cutting the employer's share of Social Security and Medicare taxes would have produced about three times as many jobs as cutting income taxes in 2011. Cutting payroll taxes specifically for firms that add new workers would generate four times as many new jobs as cutting income taxes.[17]

Nouriel Roubini, a prominent economist from New York University's Stern School of Business, suggested a two-year payroll tax hiatus and believes that the "reduced labor costs would lead employers to hire more."[18] He favors reducing payroll taxes for both employers and employees, but proposes a formula that would give more of the tax break to the employers.† Roubini considers the job creation potential

---

* Politicians sometimes imply, and people sometimes think, that tax cuts can "pay for themselves." Unfortunately, respected economists, including respected conservative economists who support low tax rates, have shown that this isn't so. There's a fuller discussion of this in our book on the federal budget: Scott Bittle and Jean Johnson, *Where Does the Money Go? Your Guided Tour to the Federal Budget Crisis* (New York: HarperCollins, 2011), 70–73.

† The federal government embraced one version of this idea when it cut Social Security taxes paid by workers as part of the compromise to temporarily extend the Bush tax cuts in 2010. The plan cut payroll taxes for employees by 2 percentage points, reducing them from 6.2 percent to 4.2 percent beginning in 2011, in hopes that it would encourage consumer spending. But the plan didn't do anything for the employers' share, which the CBO had said would actually be more effective.

of this idea so powerful that he suggests letting other taxes go up to cover the cost.[19] And remember, the CBO gives this strategy one of its top job creation ratings too.[20]

CONGRESSIONAL BUDGET OFFICE RANKINGS OF JOBS CREATION POLICY OPTIONS FOR YEARS 2010–2015
Based on years of full-time-equivalent employment per million dollars of cost to the federal budget
Policy Option
1. Reducing employers' payroll taxes for firms that increase their payroll 2. Increasing aid to the unemployed 3. Reducing employers' payroll taxes 4. Investing in infrastructure 5. Providing aid to the states (not infrastructure) 6. Providing additional refundable tax credits for low- and middle-income people in 2011 7. Reducing employees' payroll taxes 8. Providing an additional one-time Social Security payment 9. Allowing full/partial expensing of investment costs 10. Extending higher exemptions for the alternative minimum tax 11. Reducing income taxes in 2011

SOURCE: Congressional Budget Office, "Policies for Increasing Economic Growth and Employment in 2010 and 2011," January 2010, www.cbo.gov/ftpdocs/108xx/doc10803/01-14-Employment.pdf.

In 2010, the nonpartisan Congressional Budget Office looked at eleven different ideas for attacking unemployment, such as income tax cuts, payroll tax cuts, giving aid to the states, and having the government invest in infrastructure. Since all of these ideas cost money, the CBO ranked them using two criteria: (1) their impact on unemployment and (2) how much the government would spend, or lose in tax revenue. That's the simple explanation. The CBO provides a fuller one at www.cbo.gov/ftpdocs/108xx/doc10803/01-14-Employment.pdf. But brace yourself—it's an explanation only a bona fide policy wonk could love.

The problem, of course, is that even if this basic idea works, you can't do this for too long, because you need the money collected in payroll taxes to pay Social Security and Medicare benefits. Both of these programs have serious underlying financial problems. And if you're counting on business tax cuts to spur job creation, you need to think through how to handle the possibility that businesses might use the extra money to expand abroad, to buy equipment to replace workers, or—heaven forfend (as they used to say in Shakespeare's time)—to pay more to their executives and shareholders and not really hire anyone new at all. Just as you can't force individuals to spend a tax rebate on shopping, you can't guarantee that businesses will hire more workers. A Duke University/CFO Magazine survey of 1,000 chief financial officers in mid-2010 found many U.S. businesses had cash in the bank but were reluctant to expand hiring until they were sure consumer demand would come back.[21]

One way around this dilemma is to design tax breaks (or tax credits or subsidies) that go only to businesses that hire people. The government tried this idea in the late 1970s to spur employment, and whether it worked or not—and whether this is the best way to jump-start hiring—is still in dispute. In 1977 and 1978, in the wake of a tough recession, the government gave targeted subsidies to employers that reduced the cost of hiring someone new by 20 to 25 percent.

You can see why this might be an appealing plan from an employer's point of view. Ron DeFeo is the CEO of Terex Corporation, which manufactures construction equipment, and he was one of a number of business leaders who participated in the "job summit" sponsored by the White House in late 2009. He gave this idea an enthusiastic thumbs-up, saying he would "hire 100 people right now" if he could count on government incentives to offset part of the cost.[22]

But Gary Burtless, an expert on taxes and job creation at the nonpartisan Tax Policy Center, points to the downside. Back in the 1970s, paying for the subsidies increased the federal deficit. Burtless also points out how difficult it is to design a tax cut or credit that doesn't misfire: "How do we discourage a company from laying off current workers in order to become entitled to a subsidy for the new hires who replace them? How do we structure the tax credit so employers do not receive subsidies for splitting a single well-paid, full-time job into two poorly paid, part-time jobs? And how do we accomplish all of these goals while keeping the cost of the credit manageable?"[23]

If businesses are anxious about hiring given the overall state of the economy, this approach does give them a concrete, business-smart reason to go ahead and take the plunge, and it puts the focus exactly on the problem—how do we get more people hired? But figuring out the right structure for proposals such as this can be mind-numbingly complicated and very costly—especially if you try to do it for a number of years.

HOW FAR WILL THIS GET US?

Despite the blazing rhetoric over "cutting taxes to create jobs," most of the discussion in this country now is about fairly modest changes in taxes. Some economists and experts believe that small shifts in taxes one way or the other don't really make much of a difference in whether employers hire more people or not.

Roberton Williams, a senior fellow at the Tax Policy Center, argues that "firms don't hire based on tax breaks; they hire based on demand," meaning businesses won't add workers until they believe there are more and more people who want to buy what they're selling."[24] And Steven Pearlstein believes the same is true when it comes to consumer spending and tax rates on high earners. Even though "the rich account for a disproportionate share of consumer spending," he writes, "raising their taxes by a modest amount won't alter that spending or have much of a short-term impact on the economy. The reason: Wealthy people make considerably more than they spend, and they save the rest."[25]

Bruce Bartlett is also dubious. He suggests that payroll tax cuts have a limited impact because "there's little sign that labor costs are the principal factor holding back hiring."[26] Moreover, he points out, payroll tax cuts help people who already have jobs—not the unemployed—and these cuts actually go to many people who already have very comfortable incomes. For Bartlett, "under current economic conditions, all tax cuts are essentially passive"—they do little to get the economy and hiring really moving again.[27]

There is a never-ending back-and-forth between liberals and conservatives on whether cutting taxes generates economic growth and more jobs. It makes you believe cable news shows should distribute paper bags to all their guests to head off the danger of hyperventilation as they try to get their opinions across. But based on everything we have seen in preparing this book, the actual evidence is mixed. You can find evidence that it's better for jobs to have very low taxes so that people spend and save and invest. You can also find evidence that it would be better to have somewhat higher taxes so that we can have better schools, highways, airports, and Wi-Fi that's actually available when you're not in a coffeehouse—not to mention doing something about our hemorrhaging federal debt. (In thinking this over, you might want to take a look at some of this country's jobs and taxation history as it has rolled out over the last few decades. See "Are There More Jobs When Taxes Are Low?" on page 118.)

Unfortunately for those of us who would like a watertight answer on this, there isn't one. The situation may be akin to those medical trials that try to prove that X, Y, or Z is good or bad for your heart. Since so many things factor into whether you're courting a triple bypass or not, it's almost impossible to prove or disprove whether any one thing—living on oat bran or starting each day with a swig of Armagnac—will keep you out of the ER. In the same way, there are many, many factors that determine whether employers hire, whether consumers spend, whether the economy expands, and whether people are optimistic about the future and willing to act accordingly. Taxes are only one of them. If the other factors are working against us, then there's only so much tax cuts will do.

Are There More Jobs When Taxes Are Low?

If you tune in to any candidate debate—and it can be a presidential, state, or local one—it probably won't be long before you hear the phrase "job-killing taxes" or "job-creating government investments." Candidates, like most of us, often resort to shorthand to convey more complex ideas. Some candidates and elected officials believe that lower taxes help create and preserve jobs, while others believe that government spending on education, science, energy research, or infrastructure improvements can help spur job creation. Those investments, obviously, have to be covered by taxes. (At this point, we can't just count on running a tab with our national debt anymore. Those carefree days are gone, and they should be.)

Like the Jets and the Sharks, the Hatfields and the McCoys, and the Capulets and the Montagues, the two sides probably won't agree anytime soon. But understanding the details hidden beneath the shorthand may help you think this quandary through for yourself.

Below we provide a graph showing the ups and downs of the country's unemployment rates, our economic growth rate, and the top tax rate. Remember that, as in most things economic, there is lag time to be considered. The economy may be growing, but if employers are still cautious about hiring, the unemployment rate won't drop right away. In the recessions in the early 1990s and in 2001, for example, it was about a year before jobless rates started heading downward.[28]

And of course, as we've pointed out before and as we'll no doubt point out again, other factors (such as oil shortages and housing bubbles) can affect economic growth and unemployment. Tax rates are just one of them.

But the following charts do offer some food for thought—and maybe something to ask the candidates about the next time you meet one at the mall or the county fair.

Top Federal Income Tax Rates, 1970–2010

SOURCE: Tax Policy Center.

The very top income tax rate—the so-called marginal rate—has fallen dramatically since the early 1970s.

Economic Growth, 1970–2010

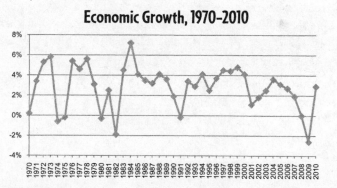

SOURCE: National Economic Accounts, March 2011, U.S. Bureau of Economic Analysis.

The annual percentage change in the gross domestic product, in constant 2005 dollars.

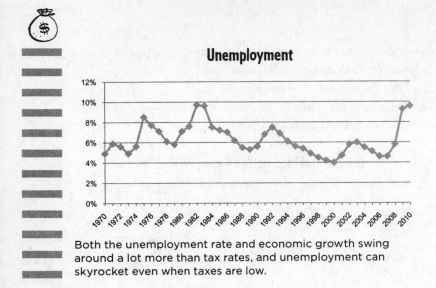

Unemployment

Both the unemployment rate and economic growth swing around a lot more than tax rates, and unemployment can skyrocket even when taxes are low.

CHAPTER 7

WOULD CUTTING BUREAUCRACY HELP CREATE JOBS?

I have wondered at times what the Ten Commandments would have looked like if Moses had run them through the U.S. Congress.[1]

—*Ronald Reagan*

Kudzu may sound like something you'd order at a sushi bar, but in fact it's an incredibly hardy vine. Back in the 1930s, the government paid people in the South to plant kudzu on the hills and alongside roads to prevent soil erosion.[2] Since erosion is a major threat to agriculture, and kudzu does hold soil in place, it seemed like a good idea at the time.

But once kudzu is planted, it moves in like an invading barbarian horde. The plant can grow thirty vines from one root, and the vines can grow up to a foot a day.[3] The proliferating greenery wreaks destruction on nearly everything in its path. It encircles trees, killing them; it blocks highway signs and can even cause power outages by damaging transmission lines. Kudzu now covers about seven million acres in the southeastern United States, and the government has changed its status from recommended cover crop to "noxious weed." It may not be the blob that ate Chicago, but it is widely known as "the vine that ate the South."[4]

If the vision of kudzu strangling innocent plants and power lines reminds you of government regulation gone haywire, you're not alone. According to a 2010 study by the Business Roundtable, one of every seven small business leaders names "government regulation" as "the single most important problem" facing their companies today. The survey respondents were four times as likely to worry about regulation as about the cost of labor.* For people in business, the issue isn't trivial.

As a matter of fact, there are a lot of people who argue that government—whether it's federal, state, or local—needs to reduce the number of rules businesses have to follow, make the rules simpler, and slash the paperwork involved. The question we take up here is whether cutting bureaucracy and red tape is a promising avenue for spurring job growth and whether it's possible to do without undercutting the protections regulation is intended to provide.

GETTING YOUR BEARINGS: A FEW FACTS TO CONSIDER

In theory, government should be able to establish some reasonable, easy-to-grasp ground rules in a few key areas (food safety, for example) without creating big government agencies and volumes of impenetrable regulations. But bureaucracies aren't new, and people have been complaining about

---

* Government regulation was the third most serious problem facing small to medium-sized businesses in the wake of the recession, coming in behind poor sales and taxes. For more see, Matthew Slaughter, "Mutual Benefits, Shared Growth," Business Roundtable, September 13, 2010, http://businessroundtable.org/uploads/studies-reports/downloads /Small_Big_Business_Report_FINAL.pdf.

them for centuries. In the 1830s, John Stuart Mill described bureaucracy in France as a "vast network of administrative tyranny."[5] Charles Dickens created the bureaucracy from hell in *Little Dorrit*. Besides being thrown into debtors' prison, his hero had to battle the dreaded "Circumlocution Office": "No public business of any kind could possibly be done at any time without the acquiescence of the Circumlocution Office," Dickens wrote. "Its finger was in the largest public pie, and in the smallest public tart."[6]

Given some two hundred years to either whittle bureaucracy down to size or bulk it up, how are we doing in the United States in the twenty-first century?

- The Code of Federal Regulations, basically the "Big Book of Every Rule of the Federal Government," is packaged in fifty volumes and boasts more than 150,000 pages.[7] In 2008, nearly 250,000 federal employees worked to administer these regulations.[8]

- The Congressional Research Service estimates that "more than 100 federal agencies issue more than 3,000 final rules" annually.[9]

- Between 1990 and 2008, spending on the agencies that oversee federal regulation more than doubled, growing about 7 percent a year.[10]

- It's not easy to pin down how much money government regulation costs businesses each year (or, as we discuss later, how much money it saves or how much pain and suffering it prevents). Expert studies that attempt to do this are often controversial.[11] One study sponsored by the U.S. Small Business Administration (itself a government agency) put the cost of federal regulation at about $1.75 trillion in 2008, or about 14 percent of the entire U.S. economy.[12] On average, that would be about $8,000 per employee.[13]

- The smaller your business is, the heavier the burden can be. When large companies have to hire personnel managers, compliance managers, lawyers, and accountants, they can benefit from economies of scale. But even though laws frequently make exemptions for businesses with fewer than fifty employees, regulations can still pack a wallop. According to the Small Business Administration, the per-employee cost of regulation for companies with more than five hundred employees averages out to just under $7,800 per worker. For a small company with fewer than twenty people, it comes out to over $10,000 per employee.[14]

- On top of federal ones, there are state and local regulations. While local rules vary (which can be a headache itself if you're trying to operate a business in many different areas), these can be costly as well. One study that examined the cost of regulation in California put the tab at about $134,000 per small business for 2007.[15] Critics say state and local regulations (and federal ones too) not only drive up operating costs, they discourage people from starting businesses and increase the investment needed to get going. If you're starting a new business in Arizona, you'll need to decide which of ten different forms to file. Is it the one that lets you "transact business in Arizona" or the one that gives you the "authority to conduct affairs in Arizona"?[16] (We're assuming that last one isn't for Tiger Woods and his assorted lady companions.)

DON'T WE NEED REGULATION?

Figuring out which regulations apply to your business and then changing what you do to comply is time-consuming, and it costs money. What's more, hardly any of us really likes being told what to do.

Still, it's worth remembering that most regulations come into being because something was going wrong somewhere. Congress passed the Sherman Anti-Trust Act when John D. Rockefeller and Standard Oil controlled more than 90 percent of U.S. oil refining.[17] In 1906, Congress enacted the Pure Food and Drug Act following publication of Upton Sinclair's stomach-turning depiction of the meat-packing industry in *The Jungle*.[18] Congress has also passed regulations to outlaw child labor, protect workers against abusive employers and discrimination, protect consumers against shifty businesses, make the financial world more transparent, and protect all of us from air pollution and contaminated drinking water. Most Americans seem to be reasonably glad that government intervened in these areas.

Plus, lack of regulation can cause problems as well. On paper, of course, no business wins by shafting its own customers. In real life, there are always people who decide that it's easier and more profitable to take ethical shortcuts. Sure, it wasn't really in the company's long-term best interest when Ford overlooked problems with the exploding Pinto in the 1970s, or Bridgestone/Firestone sold millions of defective tires in 2000, or Enron cooked the books and engaged in corrupt energy trading. But they did those things all the same.[19]

Lots of experts argue that deregulation in the financial world helped trigger both the savings-and-loan debacle in the late 1980s and the global financial crisis of 2008. (But even here, not everyone is on the same page: some say it was well-intended attempts by government to spur home ownership that fueled the mortgage crisis.)

Even so, hardly anyone in any sector of American life is happy with the scope and complexity of government regulation now. Every president since Jimmy Carter has at one time or another vowed to reduce government bureaucracy and bring some needed sanity to government regulation.[20] And when it comes to job creation specifically, two themes emerge repeatedly: government regulation (1) makes it too hard to start a new business and (2) makes hiring people too expensive.

JUST MAKE IT EASIER TO OPEN A BUSINESS

As we pointed out in Chapter 4, nearly two-thirds of new jobs in the U.S. economy are created by businesses less than five years old,[21] and studies by the World Bank and others show pretty persuasively that too much government regulation nips promising start-ups in the bud. The bottom line, according to the 2007 Global Entrepreneurship Monitor: "All other things being equal, the more onerous a country's new business regulations . . . the lower the level of ambition among a country's entrepreneurs."[22]

Compared to other countries around the world, the United States is paradise for entrepreneurs. In Suriname, it takes about two years to get a new company registered and up and running.[23] Here, it takes about six days, and you'll need to complete about six separate steps to do it. The United States is number eight out of 183 countries for the ease of getting new businesses set up. If you're an entrepreneur, there are a lot of worse places to be.[24]

The problem is that we're not necessarily headed in the right direction. Based on World Bank assessments, the difficulty of starting a business in the United States increased between 2003 and 2007,[25] and many experts believe this process needs to be streamlined. Some states, notably Hawaii, have moved to do just that. If you want to start a business in the Aloha State, you can register your new company online and complete all the required paperwork for taxes and employee hiring right on the same site.[26]

Cumbersome start-up paperwork is one way regulation stifles new enterprises, but there are others. Some regulations may inadvertently make it difficult for entrepreneurs to create new or different kinds of products. The Heritage Foundation, for one, argues that FCC regulations on radio frequencies blocked the development of wireless networks and by extension the whole world of cell phones and iPads for decades.[27] One current controversy is whether new EPA regulations on burning fossil fuels will discourage companies that want to convert coal (which, as it's currently used, releases mucho carbon dioxide into the air) into less environmentally hazardous biomass products.[28] Given the amount of money at stake, it's not unusual for established companies to lobby hard to keep these rules on the books to protect their own turf and stall the development of new products that might compete with theirs.

REIN IN HIRING COSTS

Even when you've got your company up and running, government regulation can make conducting business and employing workers more expensive than it would be if you didn't have to follow all the rules. Remember that $8,000 estimate for what regulation costs a business per employee? Here's how it shakes out.

- About $600 is for health and safety regulations (mainly OSHA, the Occupational Safety and Health Administration). This number also includes the cost of new homeland security rules put in place after September 11 (airlines keeping better track of passengers and bags, for example).
- About $800 is for tax compliance. That's for the record keeping and paperwork, not the taxes themselves.

- About $1,500 is for environmental rules.
- More than $5,000 is for what the Small Business Admin-istration calls "economic regulations."[29] This covers a whole bunch of rules on the types of products and ser-vices businesses offer and how they are priced, marketed, and sold to consumers.[30] It also includes rules on how employers treat their employees, such as the Americans with Disabilities Act and regulations on employee drug testing and collective bargaining.[31]

So does the cost (and angst) of following all these rules actually influence how many workers employers hire and how much they pay them? When businesses spend money complying with government regulations, it means they have less money left over for hiring new people and/or giving raises to employees already on the payroll. But the compa-nies need enough employees to do the work, so do the regu-lation costs really make that much of a difference?

Maybe. Here's the experience of a California farmer as chronicled by *Los Angeles Times* reporter Alana Semuels. Mike Young recently switched from growing tomatoes (which have to be picked by hand) to growing almonds (which can be shaken off trees by machines).[32] "Labor is so expensive," the farmer explained. "There's their wages, truck, insurance, workers' comp and the safety regulations." And then there's the regulation that stresses out so many California farmers in particular: having to verify workers' immigration status. Young's bottom line: "We went to a high-value crop that needed less labor input."[33] Machines are cheaper than people (and they do a nice job of picking almonds), so that certainly affected Young's decision. But the added difficulty and cost of following government regu-lations may have been the final straw.

CAN THIS IDEA SAVE THE DAY?

Most of us have had a personal run-in or two with government bureaucracy and its incomprehensible verbiage. It's there when we do our taxes. Just hiring a regular babysitter sets the bureaucracy in motion. The IRS provides pages of instructions on what you'll need to do to follow the law. In addition to paying employer taxes, you need to keep copies of "Schedule H or other employment tax forms," along with "related Forms W-2, W-3, W-4, and W-5." Of course, you'll also need to hang on to documentation backing up what you report.[34] And that's just for the feds. You'd better check on your state's employment rules and complete that paperwork too.[35]

Surely there's a better way—and we don't just mean paying the sitter under the table. Wouldn't cutting regulation and red tape be a make-everybody-happy solution? Other than tax accountants and lawyers, wouldn't Americans from coast to coast rally round this approach? Probably, but we should be clear-eyed on just how much it will actually do for job creation and what might be lost by pushing this solution too far. There are trade-offs.

First off, not everyone accepts the idea that government regulation is drowning American businesses with unacceptable costs and red tape. The Environmental Protection Agency's regulation of the chemical industry is a case in point. When the Government Accountability Office reviewed the EPA's role in regulating toxic chemicals in 2009, it recommended *strengthening* its ability to obtain information from chemical companies about the substances they use—not curtailing it. According to the GAO, only five toxic chemicals or classes of chemicals have actually been regulated since 1976 because there is such a high standard of proof for the EPA.[36]

Even so, the agency's attempts to get more information from the industry triggered formidable resistance. When the EPA proposed requiring inventory reports on chemicals used in commerce every four years rather than every five years, the pushback came quickly. As of summer 2011, the agency had suspended the proposed change in the face of opposition from the industry and some members of Congress.[37]

Moreover, some experts say studies attempting to put a dollar figure on the cost of regulation have exaggerated it.[38] Public Citizen, a nonprofit organization focusing on consumer issues, says that its examination of "thirty years of federal regulatory activity demonstrates conclusively that predictions of devastating costs have been wrong."[39] Furthermore, the group points out that the cost to business is only one side of the equation: "Far from bringing economic doom and gloom, regulatory requirements to protect the environment, workers, and consumers have often led to innovation and increased productivity. Regulation spawned many new businesses, especially companies providing hazard abatement and pollution control services. In many cases, there is no conflict between economic competitiveness and regulation."[40]

The point is that there are instances where government regulation creates jobs rather than destroys them. In this case, rolling back environmental regulations would trigger job losses in addition to other trade-offs. The impact of the 1980s Superfund legislation (to clean up toxic waste sites) is one example that has been studied. Public Citizen reports that the Environmental Protection Agency spent $7.6 billion to jump-start the cleanups, which then generated some $23.5 billion worth of business (you need equipment, chemicals, and such to do the cleanups) and more than 240,000 jobs.[41] In this case, the government helped solved a problem and spurred a little economic development on top of it.

PREVENTING BUBBLES OR CAUSING THEM?

Finally, there are the social and economic costs of leaving industries unregulated in areas that affect the welfare of us all. It's not always easy to decide where regulation is necessary and beneficial and where it veers into meddlesome obstructionism. The continuing debate on whether better regulation of the banking and mortgage industries would have prevented the housing bubble and mortgage meltdown of 2008 and 2009 is a case in point. Federal Reserve chairman Ben Bernanke, for one, has argued that "stronger regulation and supervision aimed at problems with underwriting practices and lenders' risk management" would have been an "effective and surgical approach to constraining the housing bubble."[42] If you buy Bernanke's analysis, you can only lament the wholesale pain and suffering (and job loss) that might have been prevented, even though these regulations would have cut into industry profits and reduced the number of housing sector jobs during the mortgage industry's heyday.

But not everyone agrees with Bernanke's view. The Financial Crisis Inquiry Commission was organized to examine the root causes of the housing bubble, and at least some of its members, including former CBO head Douglas Holtz-Eakin and former National Economic Council member Keith Hennessey, put the blame in a different quarter.[43] For them, government itself kicked off the housing bubble when it "subsidized and, in some cases, mandated the extension of credit to high-risk borrowers, propagating risks for financial firms, the mortgage market, taxpayers, and ultimately the financial system."[44] According to this analysis, the dreadful loss of jobs, homes, and neighborhoods could have been prevented if government had been less involved.

Even people who support robust government regulation

to protect the public often agree that what we have now is too sprawling and complicated for the country's good.[45] Public Citizen, for example, opposes loosening health and safety reporting requirements for small businesses or requiring agencies to give "amnesty" to first-time violators. At the same time, the group advocates "compliance assistance programs" wherein the government would help small businesses with the paperwork.[46] The certifiably progressive Al Gore once headed an initiative aimed at "cutting regulations and reforming and reinventing government—so that it costs less, works better, and keeps pace with today's fast-moving economies."[47]

One of the most effective ways to keep kudzu under control is to have goats munch on the stuff.[48] As much as we like the idea of having goats roam through the Capitol, statehouses, and city halls, they can't help us here. In Chapter 16 we do suggest some specific ideas for improving the way government regulation works, including ideas for having laws "sunset" or expire after a set period of time. They would fade away into the mist unless policy makers went on record saying they were still useful. At least, older regulations wouldn't be roaming the country like zombies long after they ceased to do good.

But here's the dilemma. Reducing bureaucracy and cutting red tape are surely good things—hardly anyone disputes that. The question is how much to rely on this strategy to create jobs and how to do it without putting Americans at risk. If it is done thoughtfully and carefully, cutting regulation may help. Just by itself, though, it may not be enough to do the trick. And if it is done haphazardly and carelessly, it could unravel some government protections that we really value.

YOU'RE THE BOSS—OR HOW WE GOT TO WHERE WE ARE TODAY

For those of us who don't actually run companies, it's not always apparent how complicated it can be. Besides staying on top of the competition, employers have to make sure that they're obeying all the rules about how they treat their workers, and the Department of Labor provides a handy list of relevant acts and regulations that's worth pondering. As you read it, it's easy to see why some of these rules are a good idea. It's also easy to see why employers might consider the list daunting and burdensome. Whatever you think of the list, it creates jobs for compliance officers and lawyers. There are definitely some trade-offs here.

Of course, these are only the rules employers have to follow in regard to their employees. It doesn't include the regulations they have to follow on the environment, taxes, product safety, or other arenas. And these are only the federal rules—states and municipalities often have laws on employment and worker relations. San Francisco, as just one example, has a Minimum Compensation Ordinance for any company or organization that does business with the city. It sets a minimum level of pay and stipulates that employers must provide "12 paid days off per year (or cash equivalent)" and "10 days off per year without pay per year."[49]

But back to the feds. Here, for starters, are the main federal labor laws—there are more than a dozen of them:*

---

* This list has been adapted from "Summary of the Major Laws of the Department of Labor," available at www.dol.gov/opa/aboutdol/lawsprog.htm. All quoted material is from this website.

1.   The **Fair Labor Standards Act** "prescribes standards for wages and overtime pay, which affect most private and public employment." This includes the federal minimum wage law, regulations about overtime, and restrictions on child labor.

2.   The **Employee Retirement Income Security Act (ERISA)** covers pensions and retirement policies. Some employers have to pay for insurance to protect people's pensions if a company goes out of business, for example.

3.   The **Occupational Safety and Health Act (OSHA**) covers, as you might expect, regulations protecting the health and safety of workers on the job. This is one reason construction workers wear hard hats. There's also special legislation covering safety for miners—the **Federal Mine Safety and Health Act of 1977**.

4.   The **Family and Medical Leave Act** requires "employers with 50 or more employees to give up to 12 weeks of unpaid, job-protected leave to eligible employees for the birth or adoption of a child or for the serious illness of the employee or a spouse, child or parent."

5.   The **Consumer Credit Protection Act** sets the rules employers have to follow when one of their workers is having his or her wages garnisheed because he or she owes someone money.

6.   The **Labor-Management Reporting and Disclosure Act** regulates employees' relations with unions and sets the rules for unions regarding annual financial reports, election of union officials, and so on.

7. The **Immigration and Nationality Act** covers the various visa programs that allow noncitizens to work in the United States.

8. The **Health Insurance Portability and Accountability Act** sets up rules that allow employees to keep their health insurance when they change jobs. Most of the provisions of the Patient Protection and Affordable Care Act—the Obama health care bill—affecting employers don't take effect until 2014, but after that companies with fifty or more employees will be required to offer health insurance or pay a fine.[50]

9. The **Worker Adjustment and Retraining Notification Act** "offers employees early warning of impending layoffs or plant closings," although the Department of Labor points out that it doesn't actually administer this law—it's "enforced through private action in the federal courts."

10. The **Employee Polygraph Protection Act** prevents employers from using lie detectors on their employees except in "limited circumstances."

11. The **Migrant and Seasonal Agricultural Worker Protection Act** is aimed at making sure that people who work for "agricultural employers, farm labor contractors, and associations using migrant and seasonal agricultural workers" get treated fairly.

12. There are a couple of laws covering how employers treat people who serve their country. **Veterans' Preference** regulations give people who have served in

the military a leg up in hiring for government jobs. The **Uniformed Services Employment and Reemployment Rights Act** covers employers' obligations to hold jobs for people who serve in the National Guard.

13. If your business has a contract with the federal government, say, to build a federal office building or maybe to provide uniforms for the military, you'll have to follow some special rules. It could be the **Davis-Bacon Act**, which covers construction projects, or the **McNamara-O'Hara Service Contract Act**, "which sets wage rates and other labor standards for employees of contractors furnishing services to the federal government," or maybe it's the **Walsh-Healey Public Contracts Act**, "which requires payment of minimum wages and other labor standards by contractors providing materials and supplies to the federal government."

CHAPTER 8

WOULD REVIVING MANUFACTURING HELP CREATE JOBS?

Well, we're living here in Allentown
And they're closing all the factories down
—*Billy Joel, "Allentown"*

In early June 1964, the Rolling Stones made their first visit to America, Xavier Cugat divorced his wife, and the Eastman Kodak Company had a big hit with its new line of Instamatic cameras selling for around $16.[1] Back then, Mick looked like a choirboy, Cugat was a household name,* and Kodak was a company with the golden touch. Kodak's low-cost, easy-to-use cameras (Brownies and Instamatics) changed photography from a specialist's hobby into a pursuit even the clueless could enjoy. Kodak film, in those bright yellow boxes, was one of those products a traveler could count on finding in every drugstore worldwide, from Dayton to Dubai.

---

* Cugat was a well-known bandleader who played at the Waldorf-Astoria for decades. In 1964, he divorced his fourth wife, singer Abbe Lane. Charo later became wife number five. See http://en.wikipedia.org/wiki/Xavier_Cugat.

Based in Rochester, New York, Kodak employed nearly 14 percent of the city's workforce in the early 1980s—some 60,000 workers.[2] By 2010, that number was down to 7,400.[3] Struggling to survive in recent years, Kodak has closed plants, reduced its workforce, developed new products, and phased out old ones. "Mama don't take my Kodachrome away" is the chorus to a 1973 Paul Simon hit.[4] Mama didn't have to; Kodak discontinued the film in 2009.[5]

GETTING YOUR BEARINGS: A FEW FACTS TO CONSIDER

Kodak is still hanging in there—at least as this book goes to print. But other once powerful American manufacturers are gone. Bethlehem Steel built more than a thousand ships during World War II and employed more than 160,000 workers in the 1950s, but it went into bankruptcy in 2001.[6] Blue jeans are right up there with apple pie as a classic American idea, but the last major U.S. plant making jeans closed in 2006.[7] The unavoidable truth is that American manufacturing companies have been closing and downsizing for decades, and tens of thousands of American jobs have gone with them.

In this chapter, we look at the idea of reviving American manufacturing as a way to create more and better jobs. We also lay out some of the policy changes that might be needed to do that. Here's what's been happening:

- In the 1940s, nearly one of every three American workers worked in manufacturing.[8] By 2009, it was one in ten.[9]
- Even though a smaller percentage of Americans work in manufacturing, we're still a huge player when it comes to making things. In 2008, the U.S. manufacturing sector produced $1.64 trillion worth of goods, more than the entire gross domestic product of Spain or South Korea.

Only six countries in the world produce more than the American manufacturing sector.[10]

- Productivity among American manufacturers (the amount of goods each worker produces) has soared. The increases in manufacturing have been almost twice as high as productivity increases in other sectors of the economy.

- Americans who work in manufacturing earn about 9 percent more in wages and benefits than other U.S. workers, in part because they are more likely to receive benefits such as paid leave and health insurance.[11]

- In 2008, manufactured goods made up more than half of U.S. exports. In comparison, agriculture was about 6 percent of U.S. exports.[12]

- We're not the only country where the number of jobs in manufacturing is declining. Australia, Japan, Great Britain, and Germany also have fewer jobs in manufacturing than in the past (see graph below).[13]

Employment in Manufacturing and Services, Selected Countries

Country	Manufacturing 1980	Manufacturing 2010	Services 1980	Services 2010
United States	22.1%	10.1%	67.1%	81.2%
Canada	19.1%	10.3%	67.8%	78.9%
Australia	19.4%	8.9%	64.9%	76.9%
Japan	25.0%	16.9%	54.8%	71.2%
France	26.2%	13.1%	53.0%	76.6%
Germany	34.0%	21.2%	51.9%	69.7%
Italy	26.9%	18.8%	48.8%	68.5%
Netherlands	21.3%	10.6%	65.1%	80.7%
Sweden	24.3%	12.7%	62.9%	78.4%
United Kingdom	28.3%	10.0%	61.2%	80.7%

SOURCE: Bureau of Labor Statistics, "International Comparisons of Annual Labor Force Statistics, 1970-2010."

Most Western countries have seen a shift from manufacturing to service jobs over the last several decades.

SO HOW WOULD REVIVING AMERICAN MANUFACTURING HELP ON JOBS?

Compared to other topics in this book, economists seem to be pretty much on the same page about what's behind the great disappearing act in American manufacturing jobs. There's a giant debate over how to respond, but nearly every expert we could find agrees that job losses in American manufacturing are mainly due to two unstoppable forces—advancing technology and international economic competition.[14]

As we discuss more fully in Chapters 12 and 13, both of these forces bring enormous benefits (including job creation) and worrying trade-offs (including jobs losses). Trying to put either of them back into its box is hopeless, and foolhardy as well. How could you ever un-advance technology? It's not like millions of people are going to suddenly decide to become Amish.* And what kind of lunatic would want to stop other countries from becoming more prosperous, even if it were possible? Economic growth in these countries will lift millions of human beings out of poverty, and if we play our cards right, it will provide millions more consumers for what we produce here.

So the question before us is not how to stop these forces but whether there are policy changes that would help American manufacturers flourish and compete in this new world order so that more Americans could find jobs in the sector.

FOURTEEN JOBS REPLACED BY ONE MACHINE

Let's start with technology. The truth is that companies can now produce a lot more stuff with a much smaller work-

---

* Although the pies are excellent.

force. Between 1997 and 2005, output by American manufacturing companies jumped by a stunning 60 percent.
Meanwhile, nearly four million manufacturing jobs disappeared over roughly the same period.[15]

How is that even possible without making people work
twelve hours a day and on the weekends? Well, it's the mir-
acle of rising productivity, and in manufacturing, it's been
fueled by technology.

At Kodak, for example, it used to take fourteen workers
to mix the ingredients to make film; now a machine handles
the process with more accuracy and less risk to workers.[16]
(Although the way things are going in the film business, the
machine may be out of work soon too.) U.S. automakers are
now using robots, computers, and computer-managed processes to streamline their operations. According to the Congressional Research Service, these changes have "boosted
manufacturing productivity, increased quality and cut
prices, but also led to ongoing layoffs."[17]

Despite the anguish that layoffs cause, nearly all economists see rising productivity—the ability to do more with
less—as a good thing. Companies that do not adopt better technology will eventually fail. Companies that keep
too many employees on the payroll will eventually see their
business go to leaner companies offering lower prices.
Moreover, while this process causes layoffs in one area, it
provides jobs somewhere else. After all, someone has to create, manufacture, install, and maintain the technology—
not to mention teach people how to use it.

The second undeniable reality facing American manufacturing is that foreign competition has heated up. Companies in other countries make some excellent products at
competitive prices, and this affects U.S. companies (and
jobs) in two ways. One is that some Americans buy these
excellent foreign-made products as opposed to buying

products manufactured here, so American manufacturers lose market share. Kodak once ruled the market for personal cameras. Now Japanese companies—Canon, Nikon, Panasonic—dominate the best-seller list.[18] That's assuming you're even taking pictures with a camera and not with your cell phone, none of which are actually manufactured here. The other result is that people in other countries who might buy U.S. products now sometimes choose other products instead (so U.S. manufacturers don't export as much).[19]

Like technology, foreign competition and international trade have both positive and negative effects. According to the U.S. Department of Commerce, about a quarter of domestic manufacturing jobs stem from U.S. exports— about 3.7 million jobs.[20] It's also true that foreign manufacturers who sell their product to Americans sometimes open factories here and hire American workers. Honda, Toyota, Subaru, and other foreign automakers have U.S. plants that employ some seventy thousand Americans directly.[21] That doesn't include the jobs that come from marketing, selling, and servicing Hondas, Toyotas, and Subarus. Other foreign manufacturers (Canon and Nikon, for instance) create jobs in shipping, sales, and marketing even if the products aren't made here.

International competition is unavoidable in the modern world, but whether international trade works the way it should is another question entirely, one that we take up more comprehensively in Chapter 12. While polls show that most Americans believe trade harms the U.S. economy,[22] the vast majority of economists argue for free trade, saying that it brings enormous benefits to all concerned.[23]

Of course, not every country shares our admiration for Adam Smith and his free market vision, so current trading practices fall far short of most economists' free trade ideal.* Experts are divided on the impact of today's trading situation on manufacturing specifically.[24] Federal Reserve chairman Ben Bernanke argues that trade "both destroys and creates jobs," and looking at the American economy as a whole over the last few decades, he is convinced, as are many other experts, that the United States has come out even on the jobs question.[25] According to government figures, exports of U.S. goods and services supported more than ten million jobs in 2008.[26] That's a pretty big benefit, so it might not be such a good idea to pick up our ball and go home on this one.

SO WHAT CAN WE DO?

Okay, you can't fight technology, world trade, or city hall, but that doesn't mean there aren't some thought-provoking ideas for bolstering American manufacturers and manufacturing jobs. These are some of them:

Do more to make our trading partners play fair.
Every year, the Office of the U.S. Trade Representative (essentially in charge of promoting U.S. exports) issues its National Trade Estimate, a who's-who list of countries that aren't playing fair, at least from our point of view.[27] In 2010, it included:

• India for tariffs on imported goods. The "average applied rate" is "among the highest in the world," according to the report.[28]

---

* Even back in 1776, Adam Smith was onto the free trade thing. This is how he put it: "If a foreign country can supply us with a commodity cheaper than we ourselves can make it, better buy it of them with some part of the produce of our own industry, employed in a way in which we have some advantage." Quoted in Alan S. Blinder, "Free Trade," *The Concise Encyclopedia of Economics*, www.econlib.org/library/Enc/FreeTrade.html.

- Indonesia for a drug approval process that makes it almost impossible to sell American-made pharmaceuticals there.
- China for a whole slew of policies that make it tough for U.S. manufacturers to break into the huge Chinese market.
- The European Union for subsidizing manufacture of its superjumbo aircraft, the Airbus. It makes it really hard for American aircraft makers to compete.[29]

Since there are millions and millions of potential customers for American goods in India, Indonesia, and China alone, many analysts say that lowering these barriers will boost manufacturing here. As of spring 2011, the United States had reached free trade agreements with seventeen countries ranging from Canada to Morocco to Singapore, so we can sell our products there and they can sell theirs in the United States.[30] The Obama administration has stated that this is one of its chief priorities.

Get the extra weight off U.S. manufacturers.

Many experts believe that American manufacturers are weighed down by costs that manufacturers in other countries simply don't face, and it's not just labor. According to studies by the industry, costs such as "taxation, employee benefits, tort claims, and government regulation" add almost "18 percent to the cost of doing business in the United States."[31]

Forbes magazine cuts to the chase: "You want to help domestic manufacturing? At a stroke, a 50-percent reduction in the corporate [tax] rate for manufacturers would make them more competitive with overseas facilities."[32] In fact, lowering and simplifying the corporate tax has become a bipartisan cause in recent years. President Obama has backed it, as did the 2010 National Commission on Fiscal Responsibility and Reform, which was mainly looking for ways to reduce the federal deficit.[33] The current corporate tax rate of nearly 38 percent is about twice as high as China's, at 19 percent.[34] An industry trade group, the Manufacturing Alliance for Productivity and Innovation (MAPI), has estimated that cutting the corporate tax rate by 5 percentage points would create half a million manufacturing jobs over the next decade.[35]

The snag is that current tax laws are filled with special deductions and loopholes that benefit some industries so they pay very low taxes indeed. Some of these fortunate companies like the corporate tax law just the way it is, or they want to lower the corporate tax rates and still keep their special advantages, which would put the federal budget even further in the red.[36]

Then there are the lawsuits. According to studies by the manufacturing industry, the U.S. legal system, with its class action suits, punitive damages, and other litigation-friendly procedures, makes it twice as expensive relative to GDP for manufacturers as legal systems in competing countries such as Japan, Canada, France, and the United Kingdom.[37] We discussed the cost of government regulation (federal, state, and local) in Chapter 7, but here's how Greg Bachmann, CEO of the Connecticut-based Dymax Corporation, sums up the hurdles for American manufacturers: doing business in the United States "is high-tax, high regulation, high liability . . . It's really been kind of a death by a thousand cuts."[38]

Can we get a little help here?

Most of the impetus to strengthen the manufacturing sector comes from people worried about U.S. unemployment and from communities and manufacturing companies that have been directly affected (think almost any city in Michigan). But these people have an ally in the U.S. military. For the military, the prospect that the United States might lose its manufacturing base is a potentially dangerous national security issue. During World War II, U.S. industry was called "the arsenal of democracy." The Pentagon would rather fight with American-made weapons; even more important, in a crisis you want to be able to mobilize factories at home to build what's needed.

In 2010, the Industrial College of the Armed Forces called for a "National Manufacturing Strategy" that would, among other steps, "foster a government-industry partnership to maintain a vibrant advanced manufacturing base."[39] The military's strategy includes tackling the trade and cost issues outlined above, but it also recommends doubling government funding for basic research[40] and continuing "both research dollars and early adoption incentives" in areas such as nanotechnology, biotechnology, information technology, and new materials.[41] The plan also includes steps such as protecting intellectual property rights, better job training, improving math and science education, and designating a "clearly identified institution within the Executive Branch with the wherewithal to coordinate and carry out a manufacturing strategy."[42]

The gist, at least from this point of view, is that having a strong manufacturing sector should be an important national priority, just like improving public schools or reducing crime. The Armed Forces study stresses that it is not suggesting that the government "pick winners" by backing certain companies or specific innovations. Instead, it says that a national manufacturing strategy would help all U.S. businesses thrive in a more competitive world economic climate.

HOW FAR WILL THIS TAKE US?

You could argue, and some experts do, that having fewer jobs in manufacturing is an unstoppable trend, just like advancing technology and international competition. After all, the United States isn't the only country to see jobs in this sector become a smaller part of the economy.[43] But it is also true that some countries—Germany is the notable example—have a more robust manufacturing sector than we do. About one in five Germans works in manufacturing, compared to about one in ten people here.[44]

So is developing a plan to enlarge the manufacturing sector a good idea, or is this a distraction from other steps that would be more effective at creating more and better jobs here? Critics raise a number of questions and objections, some practical and some reflecting basic principles on what government's role in the economy should be.

TRADE: SHOULD WE GIVE AS GOOD AS WE GET?

Just about every elected official you can name wants nations with unfriendly trade policies to mend their ways, but there is a mammoth divide about how to make them straighten up and fly right. To start with, there's no agreement over whether free trade agreements help or hurt us when it comes to jobs. NAFTA, the free trade agreement with Mexico and Canada, either boosted the U.S. economy or eviscerated American manufacturing, depending on who's talking—and we've been talking about NAFTA for about two decades now.

Economy in Crisis is an advocacy group that has condemned NAFTA and believes that it is "futile" to expect other countries to voluntarily adopt the free trade attitudes we endorse. "The "most obvious tool," the group says, "is tariffs on their exports." Economy in Crisis acknowledges that other countries may institute more tariffs in retaliation, but it argues that over the long run, doing more to protect our own industries would lead to "a massive incentive to renew our industrial base."[45]

Surprisingly, the idea of tough countermeasures to protect American manufacturers doesn't always come from the left. In the 1980s, President Reagan imposed "import quotas" that limited the number of cars and computer chips that could be imported from abroad (mainly from Japan at that time).[46] In the 1970s, President Nixon used voluntary import quotas and tax incentives to try to protect the American textile and shoe manufacturing industries.[47]

But to James Sherk of the Heritage Foundation, trying to protect today's U.S. manufacturers using tariffs and quotas is precisely the wrong response. Sherk says that trade isn't the chief reason U.S. manufacturing jobs have disappeared—it's technology. "These jobs have not moved overseas," he writes. "They have been automated."[48] Sherk urges Congress to resist the illusion that the United States can "bring back jobs automated by technology by restricting trade."[49]

Liberal economist and former secretary of labor Robert Reich is also dubious. Reich agrees with Sherk that the main cause for the loss of manufacturing jobs is technology, which he believes is "following the same trend as agriculture. A century ago, almost 30 percent of adult Americans worked on a farm. Nowadays, fewer than 5 percent do."[50]

To Reich, the future of the U.S. economy lies in jobs for "people who analyze, manipulate, innovate and create." Reich thinks trying to save repetitive manufacturing jobs that can easily be outsourced or done by machines is wasted effort. "Any job that's even slightly routine is disappearing from the U.S.," Reich maintains, "but this doesn't mean we are left with fewer jobs. It means only that we have fewer routine jobs, including traditional manufacturing."[51]

A JOB FOR THE GOVERNMENT?

But even if the government could find a reasonable way to bolster U.S. manufacturing against foreign competitors, is that a good role for government? Most conservatives begin to worry whenever government starts becoming an economic actor. Heritage's Sherk believes that government should take a less-is-more approach and that this strategy would help all businesses, not just manufacturers. What does Sherk advise? Holding the line on taxes, reducing government spending, reining in the deficit, reforming the legal system, and curtailing government regulation.[52]

Other U.S. administrations have taken a different approach—one that's more intentional and assertive. As we mentioned, Presidents Reagan and Nixon took direct steps to protect U.S. manufacturing, and President Clinton backed NAFTA, arguing that it would create some two hundred thousand U.S. jobs in its first two years.[53] When he signed the legislation into law in September 1993, he also said this: "When you live in a time of change, the only way to recover your security and to broaden your horizons is to adapt to the change, to embrace it, to move forward. Nothing we do—nothing we do in this great capital can change the fact that factories or information can flash across the world; that people can move money around in the blink of an eye."[54]

The Obama administration supports more government funding for research and math and science education, along with tax credits and subsidies in key areas such as green energy. The administration is also reorganizing its various departments and agencies to expedite its goal of increasing manufacturing exports and reaching additional free trade agreements.

The loss of manufacturing jobs has caused genuine and long-lasting pain in much of the country, and ignoring that reality would be callous. But how strong a hand government should take in promoting manufacturing—and that's different from helping people who have lost jobs in manufacturing find new ones—remains as widely disputed as it has been for decades. And in the end, even those who favor a more vigorous approach acknowledge that there are limits to what the government can or should do.

Ron Bloom, who served as President Obama's chief adviser on manufacturing, may be stating the obvious, but sometimes the obvious is exactly what we need to hear: "The dominant role in manufacturing will continue to be played by the private sector. . . . It is simply not feasible to make the government the principal actor in its revival."[55]

||||||||||||||||||||||||||||||||||||||||||||||||||||||||||||||||||||||||||||||||||

CHAPTER 9

WOULD IMPROVING EDUCATION HELP CREATE JOBS?

If we're number one in technology,
why do I have to call India for tech support?

—*Jay Leno*[1]

Why does Jay Leno have to call Mumbai or Bangalore for tech help? Not so long ago, if you called an American company to order something or complain, you talked to some poor soul toiling away in a call center in North Dakota or Iowa. Now it could just as easily be poor souls toiling away in call centers halfway around the world during what, for them, is the middle of the night.

WE'RE NOT IN KANSAS ANYMORE

According to the National Academy of Sciences, Mumbai-based phone banks are just the beginning. Australian radiologists read MRIs of American hospital patients. Costa Rican accountants help American accounting firms get tax returns ready by April 15.[2] GE, one of the country's oldest and most successful companies (it was included in the first

Dow Jones Industrial Average in 1896),[3] conducts most of its research and development work outside the country.[4]

As the sobering National Academy report puts it, American workers now have to compete with workers "who live just a mouse-click away in Ireland, Finland, China, India, or dozens of other nations."[5]

The flow of American jobs overseas is a complex development, one that many Americans find disturbing. One question this trend has raised is whether the United States is now being outgunned in the education and worker skills department.

There has been a raft of reports warning that unless we improve the skills of the American workforce, multinational businesses won't keep (or bring) their jobs here. If a U.S. business needs highly skilled help and can get the same quality for a better price in another country, then today's technology makes it a no-brainer to "outsource." Even state and local governments do this to save taxpayer money.[6]

Experts are divided (and a lot of Americans are very skeptical) about whether companies that send jobs offshore genuinely lack confidence in American workers' skills or are just tempted by the cheaper labor and looser regulations found overseas. (We cover this issue from other angles in Chapter 12, on globalization, and Chapter 13, on technology.) What's not in dispute is that the United States used to have the world's best-educated workforce, and that's no longer true. These days there are plenty of very well-educated people elsewhere.

Not only that, many other countries seem to be outeducating us in one particularly crucial area, the STEM fields: that's science, technology, engineering, and math. Students who don't master these subjects aren't eligible for the growing number of jobs requiring this kind of training. Without more STEM superstars than we're currently educating, the

United States may not have enough scientists, engineers, and inventors to create the innovations that lead to new companies, more exports, and more and better jobs here.

So here's our question: if the U.S. education system were better, would we be in a better position for creating more and better jobs and keeping them here?

GETTING YOUR BEARINGS: A FEW FACTS TO CONSIDER

Until recently, the United States was in the catbird seat on education. We're still near the top of the heap on fundamentals. Nearly every American gets an education, and there are places in the world where that still isn't true. (For example, the literacy rate in the United States is 99 percent; in Mali, it's 26 percent.)* But if you compare how American students do compared to students in Western Europe or some of the Asian nations whose economies are in overdrive, we do seem to be behind the curve.

President Obama, for one, has argued that improving education is critical to creating jobs. "I want you to know we have been slipping," Obama said in a 2010 education speech. "In a single generation, we've fallen from first place to twelfth place in college graduation rates for young adults. . . . [And] we know, beyond a shadow of a doubt, that countries that outeducate us today will outcompete us tomorrow."[7]

So how bad is it? The real problem doesn't seem to be *quantity* of education. It's *quality*.

• The good news is that more people complete high school. Currently, more than eight in ten Americans age twenty-

---

* UNESCO's literacy rates by country can be found at http://hdr.undp.org/en/media/HDR_2009_EN_Complete.pdf.

five and over have high school diplomas.[8] Thirty years ago, nearly half of Americans didn't get that far.[9]

- The bad news is that even with a diploma in hand, many students aren't ready for college. According to the ACT College Readiness Standards, 78 percent of students entering higher education are not adequately prepared for college-level reading, English, math, or science.[10]

- American kids also lag in science and math achievement compared to students in Western Europe and Japan. The Program for International Student Assessment (PISA) compares test scores for American fifteen-year-olds with scores for fifteen-year-olds in thirty-three other advanced nations.* In math, American students scored below students in seventeen other countries—only five countries did worse than we did. We did slightly better in science; only twelve countries had higher average scores.[11]

- About one in four Americans has at least a bachelor's degree,[12] which is a vast improvement over thirty years ago, when the figure was closer to one in ten.[13] But even though more young people start college now, dropout rates are stunningly high. Only four in ten students who start four-year college programs have graduated six years later.[14]

- We've still got the best colleges and universities on earth, right? Harvard, Yale, Caltech, MIT, and many other American colleges and universities are among the world's finest, but even here the competition is heating up. China's Peking and Tsinghua universities are rising world-class institutions, as are major universities in India and Saudi

---

* American student scores were compared to student scores in countries that are members of the Organisation for Economic Cooperation and Development. A map of countries is at www.mapsofworld.com/oecd-member-countries.htm.

Arabia.[15] Meanwhile, the National Conference of State Legislatures has concluded that "the American system is no longer the best in the world. Other countries are out-performing us."[16]

- As for producing the next generation of those job-creating scientists and engineers, the United States is number twenty-seven in the proportion of college students who complete degrees in these fields.[17] As of 2000, there were more foreign students in U.S. graduate engineering programs than there were Americans.[18]

SO HOW DOES IMPROVING EDUCATION HELP ON JOBS?

From a personal point of view, being better educated means you're more likely to earn a higher salary and less likely to lose your job during tough economic times, so everyone should definitely go as far as they can. The statistics are pretty clear on that.[19] But our focus here is on whether having a more educated population broadly will spur or support job creation.

If you confine the discussion to that, there's actually substantial disagreement among the experts. Hardly anyone sees education as irrelevant, but some analysts doubt whether improving education by itself is the most effective path to creating jobs. What's more, the phrase "improving education" can and does mean different things to different people, so this debate is more complicated than it might initially seem.

THIRTEEN WAYS TO WIN OR LOSE JOBS

When the National Academies assessed the nation's economic competitiveness, it listed thirteen criteria multinational companies use to decide where to locate new facilities.[20] Some are the usual suspects—taxes, labor costs, good transportation. But three are directly related to education: "availability and quality of innovation talent," "quality of the research universities," and "availability of a qualified workforce."

Here are their top criteria:

1. The cost of labor—how much it costs to hire employees
2. Availability and cost of the goods used to used to conduct the business
3. Availability and quality of research and innovation talent
4. Availability of a qualified workforce
5. Taxes
6. Indirect costs such as litigation and employee benefits including health care
7. Quality of research universities
8. Good transportation and communication (including language)
9. Government support for research and development
10. Reliability of the legal system in regard to property rights, contracts, patent protection, and so on
11. Growth of the domestic market—whether consumers in the country want the company's products or services
12. Quality of life for employees
13. Effectiveness of the country's economic system

SOURCE: *The National Academies,* Rising Above the Gathering Storm: Energizing and Employing America for a Brighter Future *(Washington, DC: National Academy Press, 2007).*

Let's start with "innovation talent." Not only is this something multinationals look for, but many economists see it as the open-sesame solution to getting the U.S. economy humming again. Some economic studies have suggested that most of the growth in our standard of living over the years has come from innovation and technology change.*

Here's how it works. Someone makes a discovery or comes up with an innovation, and it sets off a ripple effect through the entire economy. Successful breakthrough ideas don't just hang out in the pages of arcane research journals. They become products that have to be manufactured by companies that need workers. The best of them create entirely new markets, meaning that people will hand over good money for things they hadn't even heard of the year before.

If you're old enough to have purchased machines to play various sizes of records, 8-track tapes, audiocassettes, videocassettes, CDs, DVDs, and MP3s in your time, you're a living testament to technology's ability to create new markets. (If this list is out of date by the time we're in print, cut us some slack. Technology moves like lightning these days.) Whether you buy the latest, greatest audio technology to listen to Patti LaBelle or Purcell, you're supporting job creation in everything from manufacturing to merchandising, shipping to sales.

Even better, new products such as these can be sold to customers abroad, not just to Americans. That's important

---

* Economist Robert Solow won a Nobel Prize in 1987 for his work showing how technology and innovation contribute to economic growth and rising incomes. Read more about his work at "Robert Merton Solow," *The Concise Encyclopedia of Economics*, www.econlib.org/library/Enc/bios/Solow.html.

because it addresses one major challenge our economy is now facing: so many American consumers are in debt that they're already buying as much as they can without piling up even higher credit card bills.

If this idea makes sense to you, then one key mission is educating the STEM high achievers who set the ball rolling. If we want the breakthroughs, we need better math and science education in K-12 schools, and more young people graduating from our universities with degrees in those key STEM fields.

Maybe your mind is now circling around the "but Bill Gates dropped out of Harvard" objection. Yes, Bill Gates did drop out of college, but he was also sufficiently well educated to get into Harvard. And, as Malcolm Gladwell so effectively argues in *Outliers*, Gates spent a decade immersing himself in all things computer-related before his work began to have any impact on the world.[21]

So to have more STEM superstars, most experts agree that we'll need to provide additional incentives and support for students to pursue advanced work in these fields. We also need to have top-quality research universities armed with supergenius professors and cutting-edge technology. Remember that having top research universities was also on that National Academies "how to have a competitive economy" list.

Despite having dropped out of Harvard, Bill Gates himself strongly advocates this strategy. He urged Congress to expand National Science Foundation funding for fellowships and trainee programs. In his testimony, he pointed out that there are roughly a hundred thousand new jobs in computer science and engineering every year, but only about fifteen thousand new graduates with degrees in that field.[22]

Another priority is hiring better-trained high school science and math teachers and organizing intensive retraining for those already teaching.[23] The same goes for elementary school. Very few people become STEM high achievers without mastering the basic concepts early.

Advocates for this approach also argue that these changes need to take place in schools everywhere, not just in prosperous suburbs. If we don't improve science and math education in the schools in low-income areas too, many children, especially poor minority children, will be shut out of these fields. Not only is this unfair, it means our country loses out on the discoveries and innovations these children might make. "A mind is a terrible thing to waste" is the classic public service announcement closing line urging equal opportunity. The concept applies just as well to six-year-olds from poor families who will never have the chance to be a scientist or engineer.

AN ECONOMY THAT NEEDS MORE COLLEGE GRADUATES

We may need "innovation talent" to attract international businesses, but these companies are also looking for better-educated employees up and down the line. According to U.S. government projections, "nearly 8 out 10 new jobs will require higher education and workforce training" over the next decade.[24] That's why a whole cast of government, corporate, foundation, and education leaders are pushing for overall education reform and getting more Americans to graduate from college not just in the STEM fields but in others as well. Employers will need workers with superb analytic and communication skills and a broad knowledge of the changing world, in addition to those with scientific and technical expertise.

The College Board, which represents nearly six thousand colleges and universities and runs those delightful SAT exams,[25] has set this goal: having 55 percent of young

Americans with at least a two-year degree by 2025[26] (we're currently at about 40 percent).[27] President Obama wants every student to complete at least one year of college-level study beyond high school so that we can reclaim our position of having the world's best-educated workforce.[28]

That sounds like an admirable goal, but will it really create more jobs? The National Academies report argues that better education policies can help stem the tide of professional jobs such as accounting and radiology going overseas. Economists generally believe that a better-educated workforce works more efficiently and adapts more easily to new demands and rapidly changing technology. This then helps their companies be more successful and more likely to expand and add more positions.

The Organisation for Economic Co-operation and Development (OECD) is an international group that studies economic competitiveness in developed countries including the United States. OECD research suggests that raising student achievement (measured by international tests) is a powerful generator of economic growth. The projections for the United States are worth pondering. According to the models OECD researchers created, improving American students' scores by 25 points in twenty years (roughly a 5 percent improvement) would up U.S. economic growth by more than $40 trillion over their lifetimes.* Bringing our scores up to those of Finland would increase U.S. "GDP by $100 trillion over the lifetime of a child born in 2010."[29]

---

* In 2009, for example, American fifteen-year-olds averaged scores of 500 in reading, 487 in math, and 502 in science in the so-called PISA assessment, designed to compare students in different countries. Stuart Kerachsky, Program for International Student Assessment (PISA), 2009 Results, National Center for Education Statistics, http://nces.ed.gov/surveys/pisa/ppt/pisa2009handout.ppt.

The OECD numbers are tantalizing, and economic growth is certainly a major component of job creation. But here's something to keep in mind: U.S. job creation has lagged over the last decade even when the economy was growing nicely, so that by itself isn't the whole answer. What's more, there's the issue of how we actually go about turning our schools into world-class educators.

You could fill several libraries with books analyzing what our system does right and wrong and offering different ideas for fixing it. We can't even begin to summarize them here. The U.S. Department of Education has put out its Race to the Top agenda for change.* The College Board has a ten-point set of recommendations.[30] The American Enterprise Institute's *Education Outlook Series* offers another perspective on education reform.[31] We have listed many other resources at www.wheredidthejobsgo.org. But, assuming we could figure out the right formula, or at least an effective one, the question of job creation remains.

CAN THIS IDEA SAVE THE DAY?

Hardly anyone disputes the importance of education (we're talking adults here—not teenagers complaining about having to learn polynomials or read *Moby-Dick*). But the precise connection between a good education system and having an economy that creates and holds on to jobs is more complex.

Most experts believe it's crucial for the country to do more to cultivate top scientific and engineering talent and maintain world-class research universities and institutes, but some say this won't be a game-changer unless we actually

---

* There's an overview of this initiative at www.whitehouse.gov/the-press-office/fact-sheet-race-top.

make use of their discoveries. Vivek Wadhwa, an entrepreneur, and Robert Litan of the Kauffman Foundation say the most crippling problem is "that we've dropped the ball on translating this science into invention." We're coming up with good ideas, but we're not commercializing them.[32]

According to Wadhwa and Litan, "The vast majority of great research is languishing in filing cabinets, unable to be harnessed by the entrepreneurs and scientist-business-people who can set it free." According to their analysis, less than 1 percent of basic research ever leads to a viable product or business. That's why they argue that it's essential for top chemists, physicists, and computer science wizards to know something about marketing, finance, and product development—at least enough to envision the business possibilities and seek out partners who can get their ideas to market.[33]

Other economists question the overall premise that improving education will actually spur economic and job growth. University of Rochester economist Michael Rizzo is skeptical of studies showing a strong connection between more education and more robust economic growth. He emphasizes "the difficulty of controlling all the factors influencing growth" in studies such as these.[34] If his complaint sounds a tad familiar, it should. This is a perennial problem that comes up all the time in studies on job creation and economic growth. Whether you're talking about tax cuts, reducing government regulation, or improving education, it is extremely difficult—maybe impossible—to prove that the one component you're studying is what really caused the result.

Rizzo also cites some real-life examples that raise the question: Hong Kong saw stellar economic growth before its education level improved (although students there do quite well on international tests now). He compares that to what happened in some African countries that poured money into improving education only to see minimal or no economic benefits. "To be clear," he writes, "I am not arguing that education is not important. What this does show is that neither is it a guarantee of success, nor [is] lack of it a guarantee of failure.[35] We might also toss Argentina into the mix. Here's a country with a remarkably well-educated elite that still managed to sink itself into one of the great debt crises in history. It took the Argentines years to dig out.

Economist and columnist Paul Krugman also cautions against relying mainly on education to stop the *drip drip drip* of jobs being automated or disappearing abroad. Krugman argues that getting a BA is no longer a ticket to a high-paying job guaranteed to stay in the United States. For Krugman, if all we do is ramp up college completion, we'll end up "giving workers college degrees which may be no more than tickets to jobs that don't exist or don't pay middle-class wages."[36]

Author and career expert Marty Nemko also worries that producing too many college graduates today means that more people with bachelor's degrees will be working at jobs that don't require them. The government projects that most new jobs will require college-level training, but that's "new" jobs—not every single job in the economy. For Nemko, the balance right now is out of kilter. "We now send 70 percent of high-school graduates to college, up from 40 percent in 1970. At the same time, employers are accelerating their offshoring, part-timing, and temping of as many white-collar jobs as possible. . . . Meanwhile, there's a shortage of tradespeople to take the Obama infrastructure-rebuilding jobs. And you and I have a hard time getting a reliable plumber even if we're willing to pay $80 an hour—more than many professors make."[37]

DO ALL GOOD JOBS REQUIRE A COLLEGE DEGREE?

Nemko's argument also raises the question of whether we're putting too much emphasis on attending college and underrating the job growth potential in the trades and in other fields that simply don't require a liberal arts education or an advanced science or math degree. The Bureau of Labor Statistics prepares an *Occupational Outlook Handbook* that describes dozens of different fields and forecasts how many jobs will be available down the line. Here's what it says about "aircraft and avionics equipment mechanics and service technicians":

> Also contributing to favorable future job opportunities for mechanics is the long-term trend toward fewer students entering technical schools to learn skilled maintenance and repair trades. Many of the students who have the ability and aptitude to work on planes are choosing to go to college, work in computer-related fields, or go into other repair and maintenance occupations with better working conditions. If this trend continues, the supply of trained aviation mechanics may not keep up with the needs of the air transportation industry.[38]

And here's what it says about ironworkers:

> Employment of structural and reinforcing iron and metal workers is expected to grow 12 percent between 2008 and 2018, about as fast as the average for all occupations. The rehabilitation, maintenance, and replacement of a growing number of older buildings, power plants, highways, and bridges also are expected to create employment opportu-

nities. . . . a lack of qualified applicants challenges the education and retraining needs of the industry to meet the demands of employment growth.[39]

The point is that the country will still have jobs that require training but don't require college degrees, and we may not have enough qualified people to fill them. The government does support some job-training programs that are designed to fill some of these gaps, and most experts say they're particularly critical for people who are being displaced because of deeper shifts in the economy. But the impact of the government's investment here is not clear. In looking at the federal government, the Government Accountability Office reported that there were no fewer than forty-seven programs spread over nine agencies in 2009, spending a total of $18 billion. While nearly all those programs track how many people they place in jobs, only five have done careful studies to see if the programs were the reason, or whether people got jobs because of some other cause.[40]

There's also the question of whether just sending more people to college—at least as it has traditionally been defined—is really the best way to ensure that workers have the skills companies are seeking. Maybe the country needs more alternatives to traditional college that teach more specific job-related skills. The National Association of Manufacturers thinks so. In 2011, it introduced the Manufacturing Skills Certification Program, which aims to train half a million skilled factory workers over the next five years. The certification program will be offered in local community colleges, but it hasn't been developed by professors. It was designed by a consortium of groups that includes the Manufacturing Institute, American Welding Society, National Institute of Metalworking Skills, and Society of Manufacturing Engineers.[41]

And just because there will be more job opportunities for workers with math and science backgrounds, that doesn't mean everyone wants to do that kind of work. In one Public Agenda survey conducted in 2006, more than four in ten high school students said they would be "very unhappy" in a career that required a lot of math and science. Part of what's needed may be better teaching in these subjects so that students discover some of the excitement and creativity imbedded in science and math careers. Part of what's needed here may also be a cultural shift, so that science and engineering careers are cool, not nerdy. But even if that shift happens, some people are still better suited (and happier) working with words, pictures, or their hands than with numbers or test tubes.

And finally, there is the issue of how long improving our education system and vastly expanding college degree attainment would take to work, even if it does spur job creation. Ramping up learning in K-12 schools and having more people get college degrees might be good investments for the future, but there's a long lead time.

FROM LENO TO VONNEGUT

There are many, many reasons for countries to educate their citizens. We went for Jay Leno's complaint about having to call India for tech help to open the chapter, but we were tempted by this one from Kurt Vonnegut too: "True terror is to wake up one morning and discover that your high school class is running the country."[42]

Schools don't exist solely to produce future workers, although that certainly matters. They also produce citizens. Thomas Jefferson very famously made the point: "If a nation expects to be ignorant and free, in a state of civilization, it expects what never was and never will be."

Most Americans also see education as essential to giving all Americans an equal chance in life, and they care about that. It's how we preserve civilization as well. We can pass the knowledge of past generations on to generations to come—history, literature, art, music, science, technology, you name it.

But believing in the importance of education is not the same as counting on improved education to create jobs. There is wide agreement across the political spectrum that our schools could and should be better. There's less consensus that having better schools and more young Americans completing college will have any near-term effect on jobs, or that improving education will, by itself, lift us out of our current jobs predicament.

Some leaders—and you can probably count President Obama and Bill Gates as the most visible proponents—see improving our educational system at every level as one of our two or three top priorities, a strategy we simply can't afford to ignore. But others say that improving education without making other moves that connect more directly to near-term job creation is bound to be a disappointment.

CHAPTER 10

WOULD A MAJOR NATIONAL INFRASTRUCTURE PROJECT HELP CREATE JOBS?

Our national flower is the concrete cloverleaf.

—*Lewis Mumford*

While you're standing in the airport security line, taking off your shoes and wondering whether you'll make it to your gate on time, you might not care much about how La Guardia Airport got built. Even so, it is worth pondering. Formerly called North Beach Airport, the initial construction was completed between 1937 and 1939 through the Works Progress Administration (WPA), basically a federal jobs program.[1]

There are several curious elements here. The first is that *beach* is not the first word that springs to mind looking at the crowded, industrialized area surrounding La Guardia now. Then there's the idea that people working for the government built an airport in two years.

New York mayor Fiorello LaGuardia was a fan of the

WPA even before his name got attached to the airport.* At the opening ceremony, he praised the workers who built the airport: "Here is the living answer of the industry of these men who found themselves unemployed through no fault of their own. Here is one of the greatest monuments to the industry and skill of American labor."[2]

During the Great Depression, with peak unemployment close to 25 percent, the Roosevelt administration launched several programs that hired people to build and refurbish roads, bridges, parks, municipal buildings, and the like.[3] First about four million people got temporary jobs through the Civil Works Administration.[4] The later and more ambitious Works Progress Administration operated from 1935 to 1943, employing 8.5 million people doing everything from construction to giving music and dance lessons.[5]

Other WPA projects achieving fame and glory include the Grand Coulee Dam and the San Francisco Bay Bridge. There are also plenty of useful but less famous public buildings, post offices, schools, parks and other facilities that still exist around the country. (Next time you're in an older government building, have a look at the dedication plaque—you might be surprised at how many are WPA projects.)

Many Americans see the WPA as bold government action that kept people out of poverty and rebuilt the country's infrastructure at the same time. But even during the Depression, some pointed to waste in WPA programs—the

---

* At the opening ceremony, a plane flew overhead with a banner suggesting calling the airport "LaGuardia" (http://graphics8.nytimes .com/packages/pdf/topics/WPA/39_10_16.pdf). Today, unfortunately, the overwhelmed airport routinely ranks at the bottom of customer satisfaction surveys: http://en.wikipedia.org/wiki/LaGuardia_Airport.

word *boondoggle* actually stems from a controversy over the WPA.*

Others say that the tax money used to hire workers through the WPA would have done more good for the economy if it had been left in the private sector. Thomas DiLorenzo, an analyst for the Cato Institute, argues that even though the WPA spent billions of tax dollars, it wasn't effective in turning the economy around. "The economy remained depressed until the United States entered World War II," he writes. "There were more people enrolled in federal jobs programs in 1938 than in any other year of the depression, yet the unemployment rate was still 17.2 percent."[6]

Whatever the WPA's flaws, polls in 1936 showed that half of Americans wanted to renew the legislation that established it (with 41 percent opposed).[7] Overall, voters came down heavily in favor of President Roosevelt's approach. In 1936, he crushed opposing presidential candidate Alf Landon, carrying every state except Maine and Vermont.[8]

GETTING YOUR BEARINGS: A FEW FACTS TO CONSIDER

Today there are new calls for something akin to the WPA, and not just because of the jobs it would generate. Some of the power behind this idea stems from another national problem that is rearing its decrepit head: the country's basic infrastructure—our roads, bridges, electric grid, and such—is shopworn and badly in need of replacement and repair.

With lots of things that need fixing and too few jobs, it's not surprising that the idea of having government launch

---

* See "Boondoggling for Fun and Profit" in Scott Bittle and Jean Johnson, *Where Does the Money Go? Your Guided Tour to the Federal Budget Crisis* (New York: HarperCollins, 2011), 212–14.

major national construction projects that could employ millions for years is back on the table. Here's some background.

- The country has already used this strategy in a very small way. The 2009 "stimulus" bill contained nearly $100 billion to jump-start needed repairs and rebuilding for roads and bridges, dams, levees, rail, and other basic infrastructure.[9]
- According to the American Society of Civil Engineers (ASCE), the country needs to invest about $2.1 trillion over the next five years to bring our infrastructure up to par.[10]
- ASCE issues an annual "report card," awarding letter grades for fifteen infrastructure categories in the United States, including roads, dams, bridges, solid and hazardous waste disposal, and aviation, among others. In 2009, C+ was their highest grade in any category—and we earned it for taking out the trash (solid waste disposal). We got D's for eleven of the fifteen.[11]
- Want specific examples? According to ASCE, more than a quarter of the nation's bridges are either "structurally deficient or functionally obsolete."[12] Leaking water pipes (the big ones underground, not your faucet) "lose an estimated seven billion gallons of clean drinking water a day." Meanwhile, experts say the country's demand for water will increase over the next twenty years.[13]
- The electric grid consists of 220,000 miles of high-voltage power lines and another 5 million miles of the everyday electric wires you see in your neighborhood. Not only is the system aging, but the grid is also close to capacity in major population centers such as Southern California and the Northeast Corridor. Another problem is that the current grid was never designed to handle alternative

energy sources such as wind and solar. It's another component of the nation's infrastructure that is going to need work.[14]

- In 2009, a bipartisan commission led by two former secretaries of transportation concluded that there is "conspicuous evidence of our transportation system's steady deterioration."[15] Just in case you hadn't noticed this yourself on your morning commute, the commissioners said that there were "bottlenecks in all transport modes [that] had begun to compromise both the quality of people's lives and America's global competitiveness."[16]

For its supporters, the idea of hiring people to do all this rebuilding and revamping is a no-brainer. Compared to other strategies, it's obvious how it would create jobs. The government would hire people, or put up the money so states and municipalities could hire people, to do the work.

Private investment does play a role in infrastructure, particularly when it comes to the electricity grid, which has traditionally been paid for by utility companies. But most infrastructure spending is done by governments, and only by governments. Federal money built the interstate highway system in the 1950s, for example, and no one suggests that private industry would have paid for something so massive. If private companies need better highways, tunnels, airports, and seaports, they generally don't build new ones—they move someplace where the infrastructure is better. So if you want better infrastructure, you're going to need government money. And if you have a crumbling and unreliable infrastructure, you are risking the possibility that businesses will locate elsewhere (and there is some pretty spiffy infrastructure in places such as Singapore and, increasingly, in China).

But—and it seems like there's always a but—there is the

daunting question of where to get the money. By now, we are assuming that you're well aware of the country's monster debt problem. You can check on how huge it is this very day by going to the Debt to the Penny section of www .treasurydirect.gov. State and local governments, who usually have to put up some of the money, are strapped too, and sometimes have pushed back against projects even with federal backing. In 2010, for example, New Jersey governor Chris Christie canceled a badly needed $8.7 billion rail tunnel between New York and New Jersey, saying that even with $3 billion in federal money, the cost was too high (particularly if the project went over budget).[17]

Yet even if government budgets were in better shape, there are legitimate questions about whether using the money this way is the best idea. Maybe it should be targeted to education and job training instead. Maybe it would be better to leave the money in the hands of consumers, investors, and entrepreneurs, where it might be more likely to create private sector jobs.

HOW WOULD A MAJOR NATIONAL INFRASTRUCTURE PROJECT WORK?

It's probably helpful to go into a little more detail on some of the ideas being put forward and how they would be different from the 2009 stimulus projects. The plan here would be to launch large-scale, multiyear projects requiring significant government investment—projects that would both solve a particular problem *and* provide jobs. In contrast, the 2009 stimulus was designed to be a short-term remedy for getting some people back to work when the private sector, particularly the construction business, was in free fall. The stimulus laid out some general categories (energy efficiency and high-speed rail, for example), but states and localities could use the money for projects that suited their own needs. Since the emphasis was on creating jobs quickly, one chief criterion for the stimulus was that the project be "shovel-ready." It didn't matter so much what it was as long as it was ready to go.

That's not what advocates of an infrastructure program, such as Michael Lind of the New America Foundation, have in mind. He says it time for the country "to think big, for a change," and he envisions "an infrastructure program on the colossal scale . . . funded by federal debt with projects chosen by a national infrastructure bank."[18]* Citing the civil engineers' assessment we mentioned earlier, he argues that the United States needs to make major investments to get its infrastructure up to world-class levels—about $400 billion a year.

It's an awe-inspiring figure, to be sure, but the costs would be shared by federal, state, and local governments, and the private sector in some cases. Just for comparison's sake, the Treasury lost more than $500 billion when the Bush income tax cuts were extended for two years in late 2010. Lind himself opposes raising taxes before the economy has put the 2008–9 recession well behind it, but he does believe we'll have to raise taxes later to cover the infrastructure spending, probably through some sort of national sales tax or a value-added tax like those in Europe.

How many jobs could an endeavor like this create? According to researchers at the University of Massachusetts's Political Economy Research Institute, even a plan that's considerably less ambitious would generate about 1 million new construction jobs and 252,000 new jobs in

---

* The idea of having a national infrastructure bank isn't a new one. Read more about how it would work at http://en.wikipedia.org/wiki/National_Infrastructure_Reinvestment_Bank.

manufacturing.[19] The UMass researchers scoped out a plan calling for about $93 billion a year from federal, state, and local government, combined with about $55 billion a year from the private sector. (For instance, the utility industry would be expected to pay part of the costs for a new energy grid.)[20]

What does that cover? It would mean spending about $13 billion for new bridges and $17.5 billion to repair old ones, along with a $10.6 billion annual investment in the railroads. The biggie is $33 billion a year for the electric grid. It's a lot of work, a lot of jobs, and a lot of government spending.

For its advocates, this strategy has a two-pronged appeal. It modernizes our infrastructure (which is crucial for the economy, not to mention our safety and convenience), and it unquestionably creates jobs. That's what makes it so persuasive to economist Paul Krugman, who contrasts it with "measures like general tax cuts that, at best, lead only indirectly to job creation, with many possible disconnects along the way."[21] A better idea, Krugman writes, is for "the federal government [to] provide jobs by . . . providing jobs."[22]

HOW FAR WILL THIS TAKE US?

We'll return to the infrastructure theme in Chapter 16, but as you might expect, not everyone is thrilled by this idea. More than any other approach we've discussed so far, this one brings competing economic visions to the surface and raises basic questions about what the federal government can and should do to solve the country's jobs problems.

There is a heated "my statistics are better than yours" battle among the experts over the impact of the smaller-scale infrastructure projects in the 2009 stimulus package[23] and whether they created a lot of jobs or not. There's little doubt that some states were very slow getting started on their construction projects, even though they had thousands of unemployed construction workers after the housing bubble burst.[24] There just weren't as many shovel-ready projects as the Obama administration had hoped. For conservatives, this is another piece of evidence showing how poorly government often performs even when its intentions are good. But not everyone is so ready to bash the stimulus. Remember, the Congressional Budget Office says the stimulus created between 1.4 and 3.6 million jobs.[25]

Beyond the numbers tit-for-tat, there is a dispute at a more basic level. Cato's Tad DeHaven acknowledges that the stimulus spending created some jobs, but he points to the trade-offs: "The real question is whether it created any *net* jobs after all the negative effects of the spending and debt are taken into account. How many private sector jobs were lost or not created in the first place because of the resources diverted to the government for its job creation? How many jobs were lost or not created because of increased uncertainty about future tax increases and worries about rising debt creating another financial crisis?"[26]

These experts and others believe the better path is for government to step out of the way and allow job creation to flourish in the private sector. The Heritage Foundation's Ben Lieberman points to the cost of having the government promote "green jobs," which are subsidized by government spending and tax breaks.[27] In contrast, he says, the oil and natural gas industries could generate 113,000 to 160,000 new jobs by 2030 if they weren't so constrained by regulations such as restrictions on drilling.[28*]

---

* For more on the green jobs controversy, see Scott Bittle and Jean Johnson, *Who Turned Out the Lights? Your Guided Tour to the Energy Crisis* (New York: HarperCollins, 2009), 243–46.

THE PROMISE AND PERIL

Even if you can't stand flying into La Guardia and are some-how irritable enough to hate the sight of the San Francisco Bay Bridge (and that's hard to imagine), the bottom line is that by building them, the WPA gave jobs to construc-tion workers who needed them and left key infrastructure improvements behind as well. One of the best arguments for the WPA is that we're still *using* all this stuff our grandpar-ents and great-grandparents built during the Depression.

Moreover, the United States needs to upgrade its sag-ging, outdated infrastructure before it collapses, no mat-ter what. There's the public safety question of making sure bridges stay up and lights stay on. There's also the question of how many jobs may never materialize because we don't have the infrastructure we need, or because we've missed out on long-term trends (such as green energy). That's what makes this idea so forceful.

But everything's got to be paid for in this life. Where to get the money for infrastructure and whether that money would do more for the economy left in private hands—well, to use Hamlet's famous line, that is the question.

From Post Office Murals to the Merritt Parkway

Economists and historians are still debating whether the jobs programs launched by President Franklin D. Roosevelt during the Great Depression were the best way to revive the economy and get people back to work. What's less debatable is that the enterprise created a lot of buildings and other structures that are still serviceable some seventy years later and still have fans and admirers. All told, the New Deal's Works Progress Administration employed more than 8.5 million people, who built 651,087 miles of roads, streets, and highways and built, repaired, or refurbished 124,031 bridges, 125,110 public buildings, 8,192 parks, and 853 airports.[29] Here's a sample:

The New York metro area got two of its three major airports out of the WPA: La Guardia (shown here when it was built) and Newark Liberty International.

Library of Congress.

The Grand Coulee Dam was one of the largest public works projects of its era, and it's still producing electrical power in the Northwest.

U.S. Bureau of Reclamation.

The WPA didn't just build buildings. Branches of the WPA also supported the arts, backing writers, photographers, music, theatrical productions—and posters to promote them. Up-and-coming talents including Orson Welles and James Agee participated in WPA projects during the 1930s.

Library of Congress.

The famous San Francisco Bay Bridge was debated in the Bay Area for decades, until New Deal money was put behind the project.

Library of Congress.

CHAPTER 11

WOULD CLOSING THE GAP BETWEEN RICH AND POOR HELP CREATE JOBS?

Money is better than poverty, if only for financial reasons.[1]

Woody Allen

Before Britain's Prince William married Kate Middleton, he announced that he and his new bride wanted to set up housekeeping without servants. The plan didn't work out in real life. After the wedding it was determined that the couple would live at Kensington Palace, in part to help fend off the paparazzi. But people loved the idea of the prince and his bride taking care of themselves—just the way the rest of us do.

At the time, the press made some pretty pointed comparisons between Prince William and his father, Prince Charles, who has a staff of 149, including "butlers, chauffeurs, valets and chefs [who] are paid a total of £6.3 million" yearly (about $10 million).[2] At one point, rumors circled that Prince Charles has four people to help him get dressed in the morning, including someone who puts toothpaste on his toothbrush for him.[3] Prince Charles is seen as "extravagant"; Prince William was applauded for wanting to live more simply.

If it's true, the toothpaste thing does seem a bit much, and having four dressers is almost embarrassing for someone whose ancestor popularized a new way of tying a necktie.* But there is a benefit: Prince Charles is creating jobs.

Many people are appalled at the idea of any human being having enough money to pay people to help him or her get dressed. Others are envious, and a fair number of us just want in on the details. But for a lot of Americans, the issue isn't so much that some people are wealthy. It's that some people have been getting richer and richer, while the majority of American families are barely keeping ahead of inflation. In fact, that's basically what's been happening in the United States for about three decades.[4]

GETTING YOUR BEARINGS: A FEW FACTS TO CONSIDER

Even before the Great Recession, 78 percent of Americans said that the gap between the rich and the poor in the United States was too wide,[5] and there is now a full-throated debate about why this is happening and whether anything can or should be done about it. There are moral arguments in favor of limiting the gap, just as there are moral arguments for letting people keep the money they earn rather than having government "redistribute" it.

There are also, perhaps, competing visions of the country. Do we mainly envision the United States as a country with a large middle class and relatively few people living in great wealth or great poverty? Or do we mainly envision the

---

* The Windsor knot, which is a royal pain to tie: www.tieknot.com/windsor-knot.html.

country as one where people can pursue their own dreams and keep what they earn if those dreams materialize? No doubt a lot of us want it both ways.

The social and political implications of the income gap are enormous. For our purposes, though, we're going to take a fairly narrow look at the income gap, concentrating mainly on its impact on jobs.* So let's pose the question this way: if we want to create and preserve jobs—and foster jobs that pay enough to live a decent life—would adopting policies to reduce the gap between the rich and the poor be helpful, or would they be more likely to kill the goose that lays the golden egg?

Here's some background:

- The Census Bureau divides the population into fifths ("quintiles," they call them) based on income. In 2009, people in the top fifth had a median income of about $189,000; people in the bottom fifth, an income of about $15,000.[6] For the last thirty years, the incomes of the wealthiest Americans have been rising dramatically, as the graph on page 184 shows.
- Americans who earn more than $100,000 a year—that's about 20 percent of the population—collect nearly half of all income earned in the United States.[7]
- In 2009, the median income for American families over-

---

* Although we don't address them here, we recognize that the moral and social dimensions of the income gap are hugely important, as are its implications for democratic government. In fact, we hope that what's here will spur you to take a deeper look at these larger questions on your own. Here are some places to start: Barbara Ehrenreich, *Nickel and Dimed: On (Not) Getting By in America* (New York: Henry Holt, 2001); Paul Pierson and Jacob Hacker, *Winner-Take-All Politics: How Washington Made the Rich Richer—and Turned Its Back on the Middle Class* (New York: Simon and Schuster, 2010).

all was just under $50,000,[8] and more than 14 percent of Americans lived in poverty.[9]*

- That same year, the median salary for CEOs of the country's top corporations was $8.5 million.[10]
- In the 1970s, CEO pay was about 30 times the pay for the average worker. In 2009, it was 263 times as much.[11]
- Don't get too miffed, though. In 2007, the very wealthiest Americans, just 1 percent of the population, paid roughly 40 percent of all income taxes collected by the federal government.[12]
- Economists use a formula that compares and contrasts average family incomes within a country to rate whether wealth is concentrated among a small number of people at the top or spread out more evenly among the population.† Based on this measure, Namibia has the world's largest income gap, while Sweden has the smallest.[13] The United States is in the middle.[14]

---

* The government actually uses forty-eight different "thresholds" to define poverty, depending on the size and circumstances of the family. In 2009, the threshold for a person living alone was just over $11,000. For a four-person household, it was just over $22,000. The details are at www.census.gov/hhes/www/poverty/data/threshld/thresh09.html.

† If you dare, you can find a short explanation of the formula at www.cia.gov/library/publications/the-world-factbook/fields/2172.html.

Income Inequality

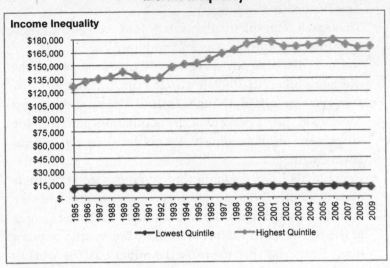

SOURCE: U.S. Census Bureau, "Historical Income Tables, September 2010."

The gap in mean household income between the richest fifth and the poorest fifth of Americans has grown over the last twenty-five years, mostly because of gains by the wealthy.

Why Is This Happening?

The pattern of the richer getting richer while everyone else financially treads water is quite clear, but the causes are less so. Like so many things economic, there's no single, definitive answer.[15] Over the past few decades, the American workforce has had winners and losers, and the widening income gap is basically the combined result. For example:

↑ **Winners: the highly skilled and very well-educated.** Since there's a shortage of highly trained, supercompetent workers in some areas, employers offer higher salaries to these people to attract and keep them.[16] Some experts believe that the reason CEO salaries have soared is because there's such a small pool of people who can lead today's huge, complex multinational corporations.[17] Not everyone buys this argument, of course; it's something we'll come back to later.

↑ **Winners: high-tech talent.** With the astonishingly rapid growth in the high-tech sector, companies that rely on top tech talent have to pay top dollar to get it. An economic analysis that looked at incomes on a county-by-county basis nationwide showed that the biggest jumps have been in regions with big tech booms—places such as Silicon Valley and King County, Washington, Microsoft's home.[18]

↓ **Losing out: low-skilled and unskilled workers.** There's an oversupply of people who may want to work hard but only qualify for jobs that don't require any particular skills. With so many people in this category, companies can keep wages low and still hire all the workers they need.[19] What's more, many economists (but certainly not all) believe that immigration has pushed wages down for lower-skilled workers. (There's more on the impact of immigration on jobs in Chapter 14.)[20]

↓ **Losing out: people who once worked in manufacturing.** Jobs in manufacturing often provide good wages for people without a college education, but as we discussed in an earlier chapter, these jobs have been disappearing. Some companies have moved their operations overseas to take advantage of cheaper labor; some have closed entirely because of foreign competition; some are using new processes that require fewer workers. Between 2000 and 2008, nearly four million manufacturing jobs disappeared.[21] The employees who used to have these jobs often have to take other jobs that frequently pay less.

↓ **Losing out: unions.** Progressive analysts often point to the decline of unions as one reason the income gap has grown. In 1983, about 20 percent of American workers were in a union; by 2010, that figure had dropped to about 12 percent.[22] When employers don't have to bargain with unions, they have more leeway to keep salaries and benefits lower.[23]

There are other possible contributing factors, such as poor public schools (which leave many workers with inadequate skills) and the expansion of pay-for-performance compensation in fields ranging from sales to classroom teaching (which allows motivated, ambitious workers to earn more than their less energetic colleagues).[24] The bottom line: the gap is due to a mix of factors, and many experts admit that they don't really understand all the causes.

SO HOW WOULD CLOSING THE INCOME GAP HELP ON JOBS?

Since there's no watertight explanation for what's causing the income gap, it's not clear that trying to close it would necessarily lead to more job creation or better wages. Still, there is no doubt that the accumulation of wealth among a small group, combined with stagnating wages for the rest of us, frustrates and disturbs most Americans. There are countless ideas about how to respond, but we'll highlight here the ones that have the most direct connection to jobs.

Executive Pay and Big Wall Street Bonuses

Public outrage over CEO pay packages and multimillion-dollar bonuses reached the boiling point in 2008 during the debate over the bank bailout. CEOs at banks, mortgage companies and Wall Street investment firms pulling in astronomical pay had driven their companies over a cliff, causing thousands of people to lose their jobs (millions if you count the broader damage to the economy). Taxpayers had to ride to the rescue. Then executives at firms such as AIG continued to pay out bonuses, even as their companies had to take bailouts to survive. Even in these circumstances, some finance industry leaders maintained that they couldn't attract top talent unless they could offer multimillion-dollar pay deals. The industry's position boggled the minds of most Americans, and many experts join the public in saying "hogwash."

The critique is more than moral indignation, although there's plenty of that. Critics charge that paying outsized salaries to CEOs and other top executives chips away at fundamental economic principles. When so much money goes to upper management, there's less left over for shareholders or workers or for the company to invest in its future. For author David Bolchover, these huge payouts are an "unwarranted plunder of shareholder funds."[25] He argues that "the myth of rare talent" is basically an invention of the financial industry itself, one that's kept alive by "the high earners themselves," along with "headhunters, business schools and management consultancies that sell their services to company chiefs."[26]

Even more dangerous, some critics say, is the short-term thinking that ultrahigh pay can encourage. If you can make a fortune to last a lifetime in just a few years, the incentive is to do whatever it takes to be "successful" in the near term. Maybe you put your efforts into mergers and acquisitions to bump up the bottom line today, even though they sap the company's strength tomorrow. Maybe you rely on layoffs and outsourcing to push up profits and stock prices quickly, rather than developing new products and services that would ensure the company's future.

Some believe that superhigh executive pay, particularly at established firms, deters entrepreneurship. With so much money to be made in finance and the corporate world, why would promising young people choose the slower, riskier path of the entrepreneur?[27] Sure, there's a Mark Zuckerberg here and there, but an awful lot of Yale and Princeton grads head for Goldman Sachs rather than starting companies of their own.

Gargantuan CEO pay may also fuel the recklessness that pushed the financial and banking sectors to the brink of collapse in 2008, according to journalist Andrew Ross Sorkin.[28] When CEOs earn so much money that their own future is ensured no matter what, they begin to see the wheeling and dealing as a game. A rash bet isn't particularly dangerous to them. If it pays off, they win. If it doesn't, their investors and employees may lose out, but they'll still have the yacht, the jet, the ranch in Montana, and the estate in Greenwich. The main thing the CEO loses is pride.

Back in the 1990s, President Clinton wanted to "end the practice of allowing companies to take unlimited tax deductions for excessive executive pay,"[29] and others have urged changes that would encourage investors and corporate boards to scrutinize executive pay more closely. And of course, we're continuing to debate higher tax rates for top earners. For those worried about the income gap, raising taxes on the wealthy accomplishes several purposes. It helps close the income gap, to be sure, but it also eases the country's budget problems and gives government the means to offer better education, health care, and other services for people who aren't anywhere near the top.

Boosting Wages at the Bottom

Not all the proposals target the rich. Some are designed to increase pay for workers in the middle and at the bottom. People who back these ideas make a moral case—there's just something wrong when CEO pay is 263 times larger than pay for the average worker. Many also believe the U.S. economy won't improve significantly unless the great bulk of Americans have more money in their pockets. As *New York Times* columnist Bob Herbert puts it: "Without ordinary Americans spending their earnings from good jobs, any hope of a meaningful, long-term recovery is doomed."[30]

Economist Robert Reich suggests a number of steps the country could have taken to counteract the trends driving the earnings gap: "Government could [give] employees more bargaining power to get higher wages, especially in industries sheltered from global competition and requiring personal service: big-box retail stores, restaurants and hotel chains, and child- and eldercare, for instance."[31] That might involve legislation to strengthen unions or to reinforce workers' rights directly. Reich has also recommended tying the minimum wage to inflation and pushing U.S. trading partners to offer better wages and worker protection in their own countries as a condition of trade deals.[32]

HOW FAR WILL THIS GET US ON THE JOBS FRONT?

Clearly, not everyone sees the income gap as a fundamental problem in and of itself. Although most analysts accept the Census Bureau data on the income gap,[33] many say these figures are incomplete and can be very misleading. First of all, they don't include all the nonmonetary benefits Americans receive from their employers (health insurance, for example) or from government (food stamps, health care).

Add these in, and the income gains for middle-class and poorer Americans aren't as pitiful as they initially appear.[34]

Second, as the Heritage Foundation's James Sherk argues, differences in income aren't nearly as important to most Americans as a rising standard of living.[35] Despite the growing gap between the rich and the poor, he points out, "most Americans today enjoy larger and better-equipped homes, better health care, more education, and more household goods than ever before."[36] In fact, with all the TVs, DVDs, computers, iPads, MP3 players, and more around the house, many Americans are beginning to protest that we have "too much stuff."[37] It's a complaint that suggests that the income gap isn't producing the impoverished lifestyle many imagine.

Many experts say that trying to limit what upper-income people earn or increasing taxes on their wealth could unleash a series of unwanted repercussions. Upper-income earners might be less likely to invest or start new ventures of their own. Some might move their money offshore. They might curb their own spending, which supports jobs of all sorts. Just ask shopkeepers in Beverly Hills, Vail, or Soho. Many Americans believe that having government insert itself into the compensation decisions of private companies—such as trying to curb CEO pay—is illegitimate and should never be pursued.

PUNISH THE RICH OR REDUCE POVERTY?

Probably the most powerful argument experts in this group make is that trying to narrow the income gap is the wrong goal. The Cato Institute's Michael Tanner writes that "income inequality is the wrong focus for government policy. After all, if we doubled the income of every American tomorrow, inequality would actually increase—but we would also lift a lot of Americans out of poverty."[38] That, Tanner says, should be the objective: "not punishing the rich, but reducing poverty."[39] The way to do that, many would argue, is to encourage economic growth, spur job creation, improve education, and ensure that people can move up the economic ladder if they have the talent and the drive.

Throughout these pages, we've returned to the same question again and again: what are the best ways to rev up the jobs engine and get the economy back to where we want it to be? We can't do everything; we have to make choices. Maybe curbing executive pay would prompt CEOs to focus more on the future. Maybe higher taxes on wealthier Americans are both fair and warranted. But if we choose to do these things, is it because they reduce the income gap? Or is it because they are good ideas that are likely to lead to good results for the job market?

Like many of the topics we've covered, there's a subtle equation here. By itself, reducing the income gap may or may not be the best path out of our jobs problem. But even if there are other, better ways to stimulate job growth, that doesn't mean that the income gap doesn't matter at all.

In fact, the income gap matters for a lot of reasons. When we were researching this chapter, we were struck by this statement from the OECD, which looks at income trends internationally: "Growing inequality is divisive. It polarizes societies, it divides regions within countries, and it carves up the world between rich and poor. Greater income inequality stifles upward mobility between generations, making it harder for talented and hard-working people to get the rewards they deserve. Ignoring increasing inequality is not an option."[40]

In the end, ignoring the income gap will probably come back to bite us. But that doesn't mean that focusing solely on reducing the gap will pay off when it comes to getting the jobs we want.

Look for the Union Label?

The 1979 film *Norma Rae* tells the story of a young woman who tries to rally her coworkers to form a union. Based on the life of North Carolina textile worker Crystal Lee Sutton,[41] the movie portrays the workers' grueling jobs—long hours in a noisy, dust-filled factory for dismal pay. *Norma Rae* used an actual textile mill as a location and real workers as extras.[42]

For many, *Norma Rae* shows exactly why workers need unions—so they can protect themselves from managers who might exploit them. A number of experts believe that strengthening unions and upping the number of workers who join is one way to support middle-class wages and mitigate the gap between the rich and the poor. Union membership has fallen over the last few decades. In 1983, about 20 percent of American workers were in unions, compared to about 12 percent now.[43]

"We shall never cease to demand more until we have received the results of our labor" is how union pioneer Samuel Gompers* summed up labor's goal early in the late nineteenth century.[44] In Gompers's era, corporate barons used violence and intimidation to try to prevent mining and manufacturing workers from unionizing. With much of the nation recoiling from these tactics, in 1935 Congress passed the National Labor Relations Act, which set up rules allowing employees to form unions and requiring employers to bargain with them.[45] In 2010, nearly fifteen million Americans belonged to unions. Employees in "education, training, and library occupations [have] the highest unionization rate"—37 percent.[46]

---

* Gompers was head of the American Federation of Labor, which later joined with the Congress of Industrial Organizations, headed by John Lewis, to form the AFL-CIO.

Are Unions Like OPEC?

Although many Americans take pride in the union movement and believe that unions are an essential counterweight
to management, others see them as a negative force—one
that has inhibited job creation and undercut American companies' ability to compete with rivals overseas. Some of the
strongest pushback relates to unions for government workers. Critics say the wages, pensions, and work rules unions
have negotiated for government employees are simply unaf-
fordable given budget pressures at the local, state, and federal level.[47]

For the Heritage Foundation's James Sherk, unions are
essentially "cartels" that attempt to control the supply and
price of labor much the way OPEC controls the supply and
price of oil.[48] By continually demanding higher wages for their
workers, Sherk argues, they drive up consumer prices and
make American-made products too expensive in the international marketplace. Union contracts also cut into profits and
undermine companies' ability to invest in new products and
services and/or expand their businesses, ultimately reducing
the number of jobs available.[49]

There's always been a tension between how businesses
define their interests and how unions define theirs. This issue
heated up considerably in early 2011 when Boeing decided
to open an assembly plant for its 787 Dreamliner aircraft in a
nonunion shop in South Carolina rather than manufacture the
plane in its Puget Sound plant, which is unionized.[50] Boeing
CEO Jim McNerney was quoted as saying that even though
there are costs associated with moving some of the company's operations to South Carolina, they "are certainly more
than overcome by strikes happening every three or four years

in Puget Sound and the very negative financial impact on the company."[51]

Union officials complained, and the National Labor Relations Board (NLRB) ruled that Boeing's decision was "inherently destructive of the rights guaranteed employees." The National Association of Manufacturers—not surprisingly, perhaps—sees the situation differently: "No company will be safe from the NLRB stepping in to second-guess its business decisions on where to expand or whom to hire" is how Joe Trauger, one of its vice presidents, responded.[52]

Disputes in the Public Sector Too

But disputes between employers and unions aren't confined to the private sector. Slightly over half of union members today work for government rather than private industry,[53] and arguments about the role of public employee unions have emerged repeatedly over the last few years as mayors and governors facing severe budget shortfalls have tried to lay off workers, curtail raises, require public employees to contribute more to their health care and retirement plans, and in some cases limit unions' ability to bargain collectively for their members.

Some officials say these changes simply must be made to avoid fiscal crises, but others see the disputes as evidence of a far-reaching attempt to short-circuit and undercut the unions' power.[54] The issue has become something of a political litmus test, with conservative elected officials typically pushing back on union demands and more progressive political leaders (including President Obama) typically defending and promoting them.

Stand and Deliver?

Views on the pros and cons of teachers' unions show how some of these arguments play out. Most Americans admire teachers, and in 2010, some six in ten Americans considered them underpaid, with only 7 percent saying that they earn too much. Yet half of the public (50 percent) also said that teachers' unions are "an obstacle that keeps schools from getting better," with only 35 percent believing that they "help make schools better."[55] Critics say unions make it hard to remove teachers who aren't up to the job and difficult to reward teachers who are especially effective. Many also believe that teachers' unions have been unreasonable in the light of state and local budget problems.

Not surprisingly, most teachers see things differently. Eight in ten say that "without collective bargaining, the working conditions and salaries of teachers would be much worse."[56] Most also worry that without a union, they would have "nowhere to turn" if they were "facing unfair charges from parents or students."[57]

Wreaking Havoc or Restoring the Middle Class?

So do unions help create a strong economy by ensuring that workers get good wages and just treatment in the workplace? Or are they "cartels" pushing up wages for their members but harming the economy overall? We'll revisit this issue in greater depth in Chapter 16, and we're sure the country will be debating the question for a considerable number of years.

Would We Have More Jobs If We Earned Less?

Given that Americans' incomes have been moving sideways over the past decade or so, the idea that maybe salaries are too high and that this is contributing to our job creation and unemployment problems may sound strange to most folks. Clearly, some people earn too much money—Charlie Sheen and the Wall Street CEOs who pushed the U.S. economy to the brink of disaster a few years ago come to mind.

But the idea that job creation slows down when it costs more to hire people and increases when employees are cheaper is classic economics. This axiom emerges repeatedly in the jobs issue—it's in debates over the minimum wage, employers' costs for health care and other benefits, and the taxes employers pay for Social Security and Medicare. In fact, many ideas for spurring hiring basically revolve around reducing employers' costs per employee.

There's also the fairly obvious fact that one major reason companies move jobs offshore is because labor is more expensive here than it is in some other countries. Even the progressive economist Paul Krugman accepts the premise in some circumstances: "Workers at any individual company may be able to save their jobs by accepting a pay cut . . . pay cuts at, say, General Motors have helped save some workers' jobs by making GM more competitive with other companies whose wage costs haven't fallen," he wrote in spring 2011.[58]

But Krugman strongly disputes the idea that lowering American wages overall would spur job creation or improve the economy, so we'll come back to his point a bit later.

Just in case you're getting really, really depressed at this point, we should assure you that hardly anyone is actually suggesting policy changes that would deliberately or specifically reduce pay for American workers.* However, there are many economists and business leaders who believe that some current policies and regulations designed to beef up people's incomes and improve their working conditions are backfiring. They believe that having a "flexible labor market" is a much better way to promote job growth and prosperity.

The U.S. Chamber of Commerce, for example, analyzed state regulations on "minimum wage and living wage laws; unemployment insurance and workers compensation; wage and hour policies; [and] collective bargaining issues," among other factors. These laws are designed to help workers and protect them from being exploited by their bosses. But from an employer's point of view, they make hiring employees and keeping them on the job more expensive. Since employers have to watch the bottom line, they may avoid adding workers or expanding their businesses if these costs are too high. The Chamber study concluded that lifting these "burdens" would be like "a free shot of economic stimulus—equal to approximately seven months of job creation at the current average rate."[59] The Chamber study also estimated that freeing entrepreneurs from these constraints would result in an additional fifty thousand new businesses every year.

As is nearly always the case in these disputes, opposing

---

* You can certainly find a few blogs suggesting this. Here's one, but even the author admits it's "whimsical": Ken Mayland, "To Create Jobs, Cut Everyone's Pay 10%," Market Watch, August 16, 2010, www.marketwatch.com/story/cut-everyones-pay-10-to-create-jobs-2010-08-16.

analysts start flinging statistical bricks at each other over the data, and it's certainly true that some experts criticize the way the Chamber of Commerce conducted its study.[*]

But for those of us who aren't keen data crunchers, the larger debate is the more important one. There are simply two schools of thought here. For the Chamber and many other experts, a flexible labor market where individual employers can decide what they need to pay to get the help they need and where they can hire and fire without too much government interference produces much better results for the economy and our standard of living over the long haul. It's in employers' self-interest to pay competitive wages. If they don't, they'll be scrambling for workers and coping with routine turnover and underqualified employees (because the ones with better skills will go elsewhere). According to this school of thought, it is better for government—state or federal—to stay out of the picture as much as possible.

But there's another argument to be made. For the Economic Policy Institute (EPI), "economic policy has not supported good jobs over the last 30 years or so. Rather, the focus has been on policies that were thought to make consumers better off through lower prices: deregulation of industries, privatization of public services, the weakening of labor standards including the minimum wage, erosion of the social safety net, expanding globalization, and the move toward fewer and

---

[*] See, for example, Ross Eisenbrey, "The Chamber of Commerce's Employment Regulation Index," Economic Policy Institute, March 16, 2011, www.epi.org/analysis_and_opinion/entry/the_chamber_of_com merces_employment_regulation_index. It argues that the Chamber of Commerce study fails to meet the most basic tests of rigor expected of any serious research report.

weaker unions."[60] The result, EPI argues, is an ever larger gap between the rich and the poor and fewer good jobs.

For progressives, the trend of government being less assertive on behalf of workers has been a recipe for sluggish growth and a poor economy—not the booming one that conservatives and business leaders generally depict. For Paul Krugman, the idea that lower wages might save jobs at a particular company when the business is looking into the abyss does not mean that lower wages overall are an asset. "There's no comparable benefit when you cut everyone's wages at the same time," Krugman argues. "In fact, across-the-board wage cuts would almost certainly reduce, not increase, employment."[61]

Why would that be? Because when people earn less money, they can't spend as much, and then businesses reduce their own workforces because they can't sell as much. What's more, in a weak economy—which is what we've seen in the wake of the recession—the whole economy is endangered because fewer people can cover their expenses and pay their debts, and some are bound to fall over the edge into bankruptcy.

For people who are open-minded and not pushing an agenda from either the left or the right, there's a complicated puzzle to think through here. It leaps out of the Chamber of Commerce study, which groups the fifty states into three categories based on the degree to which they regulate labor. In the Chamber's "top tier"—the best states for business because they tread lightly on the rules for employers—are Alabama, Florida, Georgia, Idaho, Kansas, Mississippi, North Carolina, North Dakota, Oklahoma, South Carolina, South Dakota, Tennessee, Texas, Utah, and Virginia.[62] North and South Dakota (along with Nebraska) have the lowest unemployment rates in the country (less than 5 percent in March

2011), but some of the lightly-regulated states also have very high unemployment—Mississippi, South Carolina, and Georgia had jobless rates topping 10 percent as of July 2011.[63] In other words, the states with the least burdensome regulations have both the best and worst employment results.

What's more, pay scales don't seem to tell the whole story either. While some states with high jobless rates do have fairly high wages (California with its median hourly wage of $18.21 and Rhode Island at $17.39 per hour are examples), others like Mississippi ($12.83), Alabama ($14.21), and Florida ($14.71) are comparatively low-wage states (and with light regulatory environments according to the Chamber of Commerce), yet still show some of the highest unemployment rates in the nation.[64] Given all the fireworks, it's not what you'd expect, is it?

So what can you draw from all of this? Our best advice is to take all these "just do it my way" solutions with a massive grain of salt. The important discussion the United States should be having now is not whether we could eke out some more jobs by reducing Americans' incomes. The discussion we should be having is what we need to do to get more better-paying jobs here.

STUFF HAPPENS
The Big Trends That Could Affect Jobs

In the next four chapters, we'll change things up a bit. So far, we've been looking at ideas for changing things—changing laws, changing the ways businesses or schools or government operates—and considering whether these changes would create more and better jobs. Now we'll take a look at some changes that are already happening *to* us—not changes that we can choose to make or not.

We call these next four chapters the "wild cards," because it's hard to predict exactly how these trends will shake out in the future. They all are bringing some good developments and some bad, but the important thing is we can't avoid them. Instead, we need to figure out how to cope with the cons and build on the pluses. For us, when it comes to jobs, the big wild cards are:

- **Globalization**
- **Technology**
- **Immigration**
- **The aging of the baby boomers**

CHAPTER 12

WILD CARD #1

Globalization and What It Means for Jobs

Stop the world, I want to get off!
—Leslie Bricusse and Anthony Newley

If you ask economists whether globalization is good or bad for the economy, they'll look at you as if you're nuts. No question, they'll say. It's a no-brainer: we're winning. The world in general is winning. It's a settled question.

"Other nations are best viewed not as our competitors but as our trading partners," says N. Gregory Mankiw, a Harvard economist and former chairman of the White House Council of Economic Advisers. "Partners are to be welcomed, not feared. As a general matter, their prosperity does not come at our expense."[1]

In fact, the benefits of globalization are around us every day, they argue. Daniel Ikenson, associate director of the Center for Trade Policy Studies at the Cato Institute, ticks off a list of reasons:

Most Americans enjoy the fruits of international trade and globalization every day: driving to work in vehicles containing at least some foreign content, relying on smart phones assembled abroad from parts made in multiple countries (including the United States), having more to save or spend because retailers pass on cost savings made possible by their access to thousands of foreign producers, designing and selling products that would never have been commercially viable without access to the cost efficiencies afforded by transnational production and supply chains, enjoying fresh imported produce that was once unavailable out of season, depositing bigger paychecks on account of their employers' growing sales to customers abroad, and enjoying salaries and benefits provided by employers that happen to be foreign-owned companies.[2]

And yet . . . lots of Americans just aren't buying it.

Surveys show public opinion ranging from confused to skeptical on this point. When an ABC/*Washington Post* poll asked, "Do you think the trend toward a global economy is a good thing or a bad thing for our country?" 36 percent said it was a good thing, 42 percent said it was a bad thing, and a staggering 18 percent said they didn't know.[3] (In survey research, "don't knows" above 10 percent are a sign of serious public uncertainty.) An NBC/*Wall Street Journal* poll conducted a few months earlier picked up broad skepticism about trade's impact on the country: 47 percent said free trade hurt the United States, and 23 percent said it hadn't made much difference. Meanwhile only 23 percent said that trade actually helps the economy.[4]

In February 2011, as part of its "Made in America" series, ABC News found a Texas family, the Usrys, willing to let the network take everything out of their home that wasn't made in the United States. Like a lot of people, the Usrys thought they tried to buy American.

When they came home, however, they found their living room absolutely empty, with the exception of a small vase of flowers sitting in the middle of the floor.[5] Kind of a shock, not to mention uncomfortable, trying to sit on those hardwood floors.

But as the economists would point out, that's unfair and shortsighted. We didn't see, for example, what the exercise might mean for a family living in some other country or what the Usrys might have had to pay for products that were made exclusively here in the United States. In fact, maybe the Usrys wouldn't have been able to afford all their household goods if it weren't for globalization.

And second, our focus here is on *jobs*, not *things*. The mind-bender in globalization is that just because products are made overseas doesn't mean that those products aren't creating jobs here. Yes, things that used to be made here are now made somewhere else. And some jobs that used to be done here are also now done somewhere else. But it doesn't automatically mean that all this activity isn't creating jobs here too. Here's what's been happening.

GETTING YOUR BEARINGS: A FEW FACTS TO CONSIDER

Most economists say there are three big trends driving us into a global marketplace, and only some of them have to do with government policy:

- Technology, particularly communications technology, has improved. The world is becoming both a smaller place and a more productive one. But you already knew that. The cell phone in your pocket, the computer on your desk, the ability to share files "in the cloud"—all of these things mean that many people can do their jobs from almost anywhere.

- The global labor supply has increased. It isn't just that the world's population has increased, although that's certainly true. More important, the number of people who are actually *available* as workers has increased. Let's face it, until a few decades ago, there were a lot more natural barriers to globalization. During the Cold War, it was difficult, bordering on impossible, for Western companies to operate in China or Russia, putting tens of millions of workers off-limits. Millions more in places such as India and Brazil were hampered by low education levels, trade policies, and other local barriers. Even when the barriers were lowered, they just didn't have the skills.

  As people in developing countries became more educated and less hampered by political barriers, it became practical for global companies to hire them. By some estimates, the available global workforce has increased by 50 percent.

  Don't get us wrong—this is a good thing. Huge numbers of people around the world have more freedoms and are better educated. But that does put them in direct competition with Americans for more and more jobs. At the same time, it makes many of the products we need and use more affordable for us. And as people in other countries begin to have a little more money in their pockets, they're going to want some of the products and services we make here—if we're smart about nurturing that opportunity.

- Governments have embraced free trade. Over the last two decades, government leaders around the world have

become convinced that getting rid of trade barriers benefits everybody, and they've changed their policies accordingly. The model of the European Union, where internal
economic barriers have essentially disappeared, has led
to treaties such as NAFTA, which has removed many
of the restrictions on trade between the United States,
Mexico, and Canada. The World Trade Organization has
become a global forum for settling trade disputes. You
can argue whether these policies have actually helped or
hurt (and we will, shortly). There's no question, however,
that the world trade landscape is dramatically different
from how it was even thirty years ago.

PROUDLY MADE ON PLANET EARTH

The old model of international trade is pretty straightforward: country A ships raw materials, such as steel, to country B, which uses them to make cars. The cars then get sold
to consumers, including those in country A. Country A gets
jobs from making steel, which means its workers can afford
to buy cars. Country B gets jobs from making cars, which
also includes manufacturing jobs along with jobs that are
more technically sophisticated, such as engineers, designers, and marketing. The trade arrangement creates jobs in
both countries, along with profits for the people who own
the businesses. In this case, Country B also got the advantage of better-paying, more highly skilled work.

If you look around at the furniture in your house, in
fact, you can still see this model pretty much at work. In
2007, the United States exported $593 million in wood to
China and imported $20 billion in furniture. That doesn't
mean the only American workers involved are lumberjacks,
however. Most of that furniture is made under U.S. brand
names, with Americans designing the products, plus all the

marketing and actual selling of the furniture.[6] The Chinese do the actual cutting and assembling. But clearly, this is a situation where making a table is cheaper even if it has to go thousands of miles the long way around. The downside, of course, is that the United States has lost nearly two hundred thousand jobs in furniture manufacturing over roughly the last decade.[7] This is the aspect of globalization that makes it such a fearful prospect for so many Americans.

In the end, however, most economists argue that this is better for everybody. Mankiw, for example, explains this as much like hiring someone else to shovel your walk. You're willing to pay $40 to have it done. The boy next door could spend a few hours on his Xbox, which is worth $20 of his time, but if you offer him $30, he'll shovel your walk instead. "The key here is that everyone gains from trade," Mankiw writes. "By buying something for $30 that you value at $40, you get $10 of what economists call 'consumer surplus.' Similarly, your young neighbor gets $10 of 'producer surplus,' because he earns $30 of income by incurring only $20 of cost. From an economist's perspective, there's a rational and healthy process here. Each of the parties involved pays or accepts a price that seems reasonable to them, and they all get something of value out of the deal."

Economics is all about competition, but it isn't like sports, where there has to be a winner and a loser every time. As Mankiw puts it, "an economic transaction between consenting consumers and producers typically benefits both parties. This example is not as special as it might seem. The gains from trade would be much the same if your neighbor were manufacturing a good—knitting you a scarf, for example—rather than performing a service. And it would be much the same if, instead of living next door, he was several thousand miles away, say, in Shanghai."[8]

In a globalized world, however, the system has gotten incredibly complex. A complicated consumer product could draw on parts from multiple countries before its final assembly. Let's take a look, for example, at one of the most famous American products there is: airliners. While this is a ferociously competitive business, most of the world's airline passengers still fly on planes built by only two companies: Boeing, based in Seattle, and the European consortium Airbus. In fact, Boeing is one of America's biggest exporters. But where do those planes come from?

Major Global Sourcing for the Boeing 787 Dreamliner

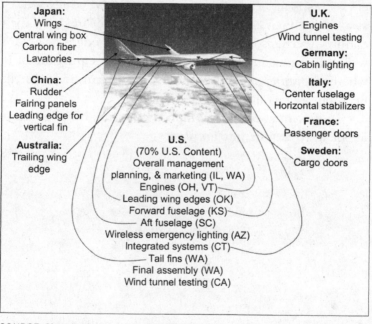

Japan:
Wings
Central wing box
Carbon fiber
Lavatories

China:
Rudder
Fairing panels
Leading edge for
vertical fin

Australia:
Trailing wing
edge

U.S.
(70% U.S. Content)
Overall management
planning, & marketing (IL, WA)
Engines (OH, VT)
Leading wing edges (OK)
Forward fuselage (KS)
Aft fuselage (SC)
Wireless emergency lighting (AZ)
Integrated systems (CT)
Tail fins (WA)
Final assembly (WA)
Wind tunnel testing (CA)

U.K.
Engines
Wind tunnel testing

Germany:
Cabin lighting

Italy:
Center fuselage
Horizontal stabilizers

France:
Passenger doors

Sweden:
Cargo doors

SOURCE: Chart by Congressional Research Service, data from "Boeing 787: A Matter of Materials," Industry Week, December 1, 2007, 35–37, plus several news articles. Photograph from the Boeing Company.

It's the global supply chain at work: Boeing's 787 Dreamliner rolls out of a hangar in Seattle, but it's still an international production.

Boeing does the final assembly for the 787 Dreamliner, and 70 percent of the plane is made in the United States. But the other 30 percent draws on manufacturers all over the world, including wings and lavatories made in Japan, a center fuselage from Italy, a rudder from China, and cabin lights made in Germany.[9] That means this flagship American product creates jobs not just here but overseas as well.

Plus, outsourcing some of that work may be shrewd business for Boeing in the long run. The fact is that Boeing can't survive on just making American planes for American airlines. Most of the world's airlines are national carriers with strong ties to their governments, who may well be swayed by having part of a Boeing product made in their country. Perhaps the fact that the leading edges of a 787's wings are made in Australia won't be the deciding factor in whether Qantas chooses to buy some Dreamliners for its fleet, but it can't hurt.

There's an even more complicated setup, however. Have a look at what goes into an iPod:

Manufacturing Supply Chain and Input Costs for the Apple iPod in 2005

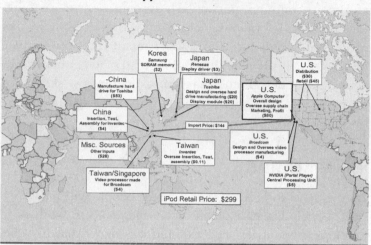

SOURCE: Chart by Congressional Research Service, data from Greg Linden, Kenneth L. Kraemer, and Jason Dedrick, "Who Captures Value in a Global Innovation System? The Case of Apple's iPod," Personal Computing Industry Center, June 2007 (and various other sources).

Just because an Apple iPod isn't manufactured in America doesn't mean it doesn't bring in money or jobs to Americans. Here's how the $299 retail price of an iPod is split up if you look at it by country.

In 2005, an iPod cost $299 retail, and a grand total of $5 worth of it was manufactured in the United States. As you can see on the chart, most of the manufacturing cost was in China, Japan, Taiwan, Korea, and Singapore. However, the cost of an iPod includes a lot more than the manufacturing. Americans designed it, managed this complicated production process, marketed it to consumers, sold it in retail stores, and took a profit: all told, that adds up to $159 of costs going to businesses and workers in the United States. About $80 went to Apple specifically.

This setup, where Americans design and sell products that are actually made in other countries, is becoming more and more common, particularly in consumer electronics. Cell phones, televisions, computers—they're actually not made here anymore. That doesn't mean they don't create jobs here. But the jobs they create tend to be either highly skilled, as in design and marketing, or very low-skill, such as retail sales. The middle-skill jobs in factories and assembly end up somewhere else.

NOTHING IS INEVITABLE

Does it have to be this way? Lots of people will tell you it does, simply because of the low wages paid to workers in developing countries. Federal statistics show that in 2006, compensation for the average American production worker was $24.59 per hour, compared to $16.02 in South Korea, $2.92 in Mexico, and an estimated *81 cents* in China.[10] You can easily look at those numbers and figure we just can't win.

But it's also true that wages aren't the only factor in a company's decision. Consider this: if workers in another country make only 10 percent of what Americans make, but they're also only 10 percent as productive, it's actually a wash as far as business is concerned. You're not actually saving any money. That's why we're not seeing jobs shifting to, say, Bangladesh or Chad, where wages may be even lower than in China, but the education levels are lower as well.

Even in better-established nations, offshoring has sometimes proved to be a disappointment, as companies wrestle with both cultural differences and quality issues. Some companies, such as Dell, moved their call centers back to the United States after customer complaints. Others find the cost savings don't live up to their original billing.[11]

Businesses also need stable governments, transparent

business practices, and reasonably honest legal systems (so, for example, you can tell if a potential partner is reliable, and whether your contract will be honored and enforced). Tax policies, health care and environmental laws, and other issues are a part of this as well. When you add up those factors, not every country qualifies as a potential trading or manufacturing partner, even if labor in the country is very cheap. As we mentioned, the World Economic Forum reported in 2010 that the United States is still ranked fourth out of 139 countries in global competitiveness.[12]

EVERY WHICH WAY BUT UP

One key aspect of globalization is that it goes both ways—or to be more accurate, it actually goes in dozens of different directions all at once. More of the furniture we buy is being manufactured abroad, but we get the advantage of the jobs in the lumber industry and jobs in the design, marketing, and sales of American furniture companies. At the same time, American consumers are buying more furniture from big international furniture companies such as Ikea and Knoll, and there are jobs for Americans in marketing and sales there. What's more, American companies are selling more of their furniture to buyers outside the United States. U.S. furniture exports jumped by 75 percent between 1990 and 1995—at the same time as so many jobs in U.S. furniture manufacturing were disappearing. The main markets are Canada and the United Kingdom, along with Japan and Mexico.[13]

And in some cases, foreign companies decide they'd rather manufacture their products here. If globalization was all bad for American workers, we wouldn't see foreign companies establishing operations here. Big-name foreign companies such as Toyota, Nokia, Seagram, and Bayer all have operations here, and in 2007 they employed 5.5 million people in the U.S. economy, exported $216 billion worth of products, and imported $533 billion in goods.[14]

SOME JOBS ARE MORE VULNERABLE THAN OTHERS

So clearly, a lot of what's going on in globalization is *specialization*. That's an idea we're kind of used to in some ways. Certain parts of the world, and certain parts of the United States, have always been associated with different kinds of businesses, and that's nothing new. You get lumber from Oregon, coal from West Virginia, television shows from Hollywood, watches from Switzerland, cheese from both Wisconsin and France. If cell phones end up being made in Taiwan, what's the difference?

Well, the difference may depend on who you are and what you're doing. Work depends a lot on local resources, including raw materials, local talent, and transportation. It always has. One reason they brew beer in Milwaukee is because lots of people immigrated there from countries where they also brew beer, such as Germany and Czechoslovakia. They had lots of people who knew a lot about making beer, along with lots of people who liked to drink it.

The difference now is that thanks to modern communications, you can tap into talent pools worldwide—depending, of course, on what you're trying to do. Not every job can be done everywhere. And it's not always the low-income, low-skilled jobs that move.

The economist Alan Blinder, for example, created a grid of which jobs might be likely to move overseas. The biggest factors aren't skills, he found, or how technical the work may be. The key questions are whether a job *has* to be done in a specific location, and whether it requires face-to-face interaction.

What Jobs Might Move Offshore?

Category	Occupation	Index number	Number of workers
I Highly offshorable	Computer programmers	100	389,090
	Telemarketers	95	400,860
	Computer systems analysts	93	492,120
	Billing and posting clerks and machine operators	90	513,020
	Bookkeeping, accounting, and auditing clerks	84	1,815,340
II Offshorable	Computer support specialists (I and II)	92/68	499,860
	Computer software engineers, applications	74	455,980
	Computer software engineers, systems software	74	320,720
	Accountants	72	591,311
	Welders, cutters, solderers, and brazers	70	358,050
	Helpers — production workers	70	528,610
	First-line supervisors/managers of production and operating workers	68	679,930
	Packaging and filling machine operators and tenders	68	396,270
	Team assemblers	65	1,242,370
	Bill and account collectors	65	431,280
	Machinists	61	368,380
	Inspectors, testers, sorters, samplers, and weighers	60	506,160
III Non-offshorable	General and operations managers	55	1,663,810
	Stock clerks and order fillers	34	1,625,430
	Shipping, receiving, and traffic clerks	29	759,910
	Sales managers	26	317,970
IV Highly non-offshorable	Business operations specialists, all other	25	916,290

SOURCE: Princeton University, CEPS Working Paper No. 142, March 2007.

Princeton economist Alan Blinder created a scale to analyze how likely it is that a particular job could be done offshore. The key factors are whether a job needs to be done in a certain location and whether it requires face-to-face contact. By that standard, most jobs stay put.

"One good bet is that many electronic service jobs will move offshore, whereas personal service jobs will not," Blinder writes. "Here are a few examples. Tax accounting is easily offshorable; onsite auditing is not. Computer programming is offshorable; computer repair is not. Architects could be endangered, but builders aren't. Lawyers who write contracts can do so at a distance and deliver them electronically; litigators who argue cases in court cannot."[15]

If you look at it that way, there's one very comforting conclusion: the vast majority of American jobs can't be sent overseas, even if you wanted to. Of the roughly 130 million jobs in America, about 100 million either can't be sent overseas at all or would be more trouble than they're worth. That includes office and retail clerks, store managers, pharmacy technicians, sound engineers, oil field roustabouts, miners, food service workers, postal employees, and actors.[16]

But roughly one-quarter of jobs could be done elsewhere, or more precisely, anywhere: computer programmers, telemarketers, data entry clerks, mathematicians, bookkeepers, graphic designers, welders, and, in a touch of irony we'll all appreciate, economists.

SO IS THERE ANYTHING WE CAN DO?

You can see the potential problem here. With globalization really rolling, if the world is one big labor market, then many of the conventional approaches won't help. There are some ideas analysts put forward to mitigate the impact for Americans—some aimed at making sure that the United States gets the good jobs, others aimed at blunting the onslaught of foreign companies selling their products here, and still others targeted at dissuading American companies from shipping their jobs offshore.

A lot of people put their faith in education: training American workers for work that can't be done overseas, and ramping up skills here so we can get the good jobs working for an American company or a foreign company doing business here. But education, in and of itself, isn't a guarantee your job won't move. The irony is that some jobs requiring significant education can be done by anyone almost anywhere in the world, while some of the jobs destined to stay in the United States don't require much education at all. For example, tens of thousands of Americans have gotten training as computer programmers in full confidence that this was a field with exploding opportunities, and they're right.

But those opportunities could just as easily be in India or Israel as here. So despite its obvious benefits, education doesn't really provide a complete answer to the impact of globalization.

Not to mention that some of economist Blinder's here-to-stay jobs don't require much education at all. The drive-through lane at McDonald's is never going to stretch all the way to Madras. But since these jobs don't require much education or training, they don't pay very well. It's pretty hard to support a family by talking through a speaker at McDonald's.

You could, of course, increase tariffs (essentially special taxes on imported goods) or impose tax penalties on companies that send jobs overseas. Higher tariffs would make imported goods more expensive here, encouraging Americans to buy our own products and giving American manufacturers with American employees a better chance at surviving and prospering. In a globalized world, however, that may not do as much good as you think. If we impose higher tariffs here, then countries such as China and India will likely respond in kind, making our products more expensive there. Someone else will get China's business instead of us. And American consumers might pay higher prices.

Another strategy is to require "fair trade," by building into trade agreements a requirement that other countries have to provide Western-style worker rights and environmental rules, or by limiting imports from countries with miserable policies at home. This would raise the standard of living in other countries, and the people there would be more able to afford American products, which could be a good thing. Arguably, it would also reduce the low-wage competitive edge that other nations have.[17]

Another element here is what critics call "currency manipulation," where a nation keeps the value of its currency artificially low to give it an advantage in trade. (If country A has a cheaper currency than country B, then it makes sense for companies in country B to ship jobs to country A, because it's cheaper.) Critics argue that China does this already. Trade agreements, with higher tariffs, could help counterbalance that.

The government could also take other steps to make outsourcing less attractive: taxing profits companies earn in foreign markets, for example, or imposing tax penalties for moving jobs. Bringing labor or professional groups into negotiations on trade deals might help too.[18] All these are aimed at increasing the costs to companies of sending jobs overseas—although, as we've pointed out, you'd need to make the penalties stiff enough to counterbalance the dramatic cost savings companies can realize when they move jobs overseas. You also have to think about the long-term trade-offs versus the short-term benefits. You may save jobs for Americans now but at the same time drive up the cost of American-made products, so over time U.S. manufacturers just aren't as competitive internationally.

Other economists argue for what they call the "Scandinavian model." In countries such as Sweden and Norway, strong unions and a government welfare state back up wages for Scandinavian workers, but businesses are also free to hire and fire freely and even outsource work. With different labor and health care policies, service jobs in retail and the food industry might be more acceptable options. Proponents say that overall, this encourages a focus on high-end work that can be done locally, while allowing businesses to churn jobs as needed in order to remain competitive and improve their productivity.[19] And from the worker's point of view, having to change jobs isn't as daunting if your health care is covered and you can count on generous transition support.

There's also the question of retraining people for the

jobs that aren't going anywhere—the ones most likely to stay here in the United States. The U.S. government has had a program for this (Trade Adjustment Assistance) since 1962, and its training provisions, in particular, have gotten good reviews. But it's also a relatively small program, and the Government Accountability Office has criticized it for being too bureaucratic.[20]

The biggest takeaway of all might be that we have to recognize that while globalization is good for more people than not, it isn't good for everybody. And since it isn't going away, we have to do something about the people who may be left behind—who are as likely to be in your IT department as on the factory floor.

As Blinder put it: "If we economists stubbornly insist on chanting 'free trade is good for you' to people who know that it is not, we will quickly become irrelevant to the public debate. Compared with that, a little apostasy should be welcome."

WHO'S BUYING AMERICAN?

There are nearly seven billion human beings on the planet, and only some three hundred million or so live in the United States. That means that there are a lot of potential customers for American-produced goods and services out there. So who is buying American stuff now? These are the ten countries that imported the most from the United States in 2010: [21]

1. Canada: $249 billion
2. Mexico: $163 billion
3. China: $92 billion
4. Japan: $60 billion
5. United Kingdom: $48 billion
6. Germany: $48 billion
7. South Korea: $39 billion
8. Brazil: $35 billion
9. The Netherlands: $35 billion
10. Singapore: $29 billion

SOURCE: www.trade.gov.

CHAPTER 13

WILD CARD #2

The Revolution in Technology and What It Means for Jobs

I, for one, welcome our new computer overlords.
—*Jeopardy champion Ken Jennings, after being
defeated by IBM's Watson computer*[1]

A lot of the writing about technology and jobs seems tinged with sadness, regret combined with a depressed acceptance of the march of progress. The human worker is always being cast as John Henry, the steel-driving railroad man in the folk story who took on the steam hammer and won, but died in the process.[2]

Consider, for example, the decline of the movie projectionist, as reported by *Slate* in late 2010. Being a projectionist was a well-paid, skilled job for a very long time, and it wasn't uncommon to meet families who'd been in the business for three generations. But thanks to digital technologies, fewer and fewer people are needed to actually run projectors. A multiplex may only need one projectionist to handle ten screens now—and an all-digital theater may not need a projectionist at all. The membership of the New York

City projectionists' union has fallen from more than three thousand members in the 1950s to only about four hundred today. "With projectionists gone, another part of our lives will lose the human touch," *Slate* mourned.[3]

Very true, very sad, and very wrenching for the projectionists who are put out of work by new technology. But there's another side to this. Being a movie projectionist is a job that exists only because a new technology was developed. There weren't any movie projectionists in 1890 because there were no movies. All through the twentieth century, that new technology of moviemaking not only grew but evolved and kept creating jobs all along the way, even as old ones fell by the wayside. The movie theater industry suffered a number of bad periods, with particular challenges in the 1950s (from television) and the 1980s (from home video). But jobs created by television and home video compensated for the losses in traditional movies.

If you look at it more broadly, you can see this at work in a report by the IBISWorld research group on the industries most in decline in the United States.[4]

Ten Key Dying Industries

Industry	Revenue 2010 (millions)	Decline 2000-10	Forecast Decline 2010-16
Wired telecommunications carriers	$154,096	-54.9%	-37.1%
Mills	$54,645	-50.2%	-10.0%
Newspaper publishing	$40,726	-35.9%	-18.8%
Apparel manufacturing	$12,800	-77.1%	-8.5%
DVD, game and video rental	$7,839	-35.7%	-19.3%
Manufactured home dealers	$4,538	-73.7%	-62.0%
Video postproduction services	$4,276	-24.9%	-10.7%
Record stores	$1,804	-76.3%	-39.7%
Photofinishing	$1,603	-69.1%	-39.1%
Formal wear and costume rental	$736	-35.0%	-14.6%

SOURCE: IBISWorld, "Ten Key Industries That Will Decline, Even after the Economy Revives," May 2011.

You can see that some (but not all) of the industries in trouble are struggling with technological change.

Now, mobile home dealers and costume rental stores aren't fields where there's been a technology boom, but just look at some of the others. Obviously, wired landline phones are taking a big hit from cell phones. The number of Americans who are cell-phone-only more than doubled between 2006 and 2010, with about one in four Americans using cell phones exclusively.[5] Newspapers are being undercut by the Internet. Home video rental, video postproduction, record stores, and photofinishing are all suffering from the switchover from physical media to digital formats. Our entertainment options are transforming from film, DVDs, and CDs to streaming video on demand, digital cameras, and iPods.

But just like the movie projectionists, most of these dying industries boomed before their bust—they were created by technology as much as they are being killed by it.

You wouldn't have telecommunications without a telephone, and if jobs lost at Verizon are replaced by new jobs at Verizon Wireless, there's not that much impact on the economy as a whole. You have to go back further with newspapers, but the advent of high-speed presses and telegraphy in the nineteenth century is what made newspapers the very first mass medium.

Sometimes the cycle of technological boom and bust seems to fly by so fast that some consumers have trouble just keeping up. The development of the VCR in the late 1970s, followed by the DVD player, created the market for home video rental. That ability to watch whatever you wanted, whenever you wanted, in the privacy of your own home led to a video store in every town, not to mention an unexpected boom in other industries, such as pornography. Now, streaming video is providing the same rental service that you used to get from Blockbuster, but from the comfort of your own couch, while adult entertainment is actually suffering a fate more like that of newspapers (there's too much competition from free alternatives—really).[6]

Meanwhile, people are being employed in the industries that are replacing these older businesses. People are manufacturing digital cameras, video editing software, wireless routers. We're losing video store clerks but gaining broadband installers. In fact, if you've tried to get any help installing broadband lately, you could be forgiven for thinking there's a bit of a labor shortage there.

You can see this at work in the *Occupational Outlook Handbook*, the federal government's official best estimate of employment trends. Just to take one example, the Bureau of Labor Statistics estimates employment in the information sector will increase 4 percent by 2018, adding 118,100 jobs. But within that category, the churn driven by technology is huge.

- Jobs in data processing and Web hosting are projected to grow by 53 percent, as more and more publishing and broadcasting moves onto the Internet. Software publishing is projected to grow by 30 percent.
- Traditional publishing including newspapers, on the other hand, is expected to decline by 5 percent, for exactly the same reason, plus a trend toward using more freelancers.
- Telecommunications jobs are projected to decline by 9 percent, even though demand will be up, because technology is providing these services more efficiently.[7]

Economists, on the whole, aren't scared of technology. Some argue, in fact, that most of the job and economic growth of the last century has been driven by technological advances. Electricity, automobiles, aviation, broadcasting, telecommunications—you can see how these have created huge numbers of jobs. Whatever has been lost economically during the past century has been more than made up for by new advances.

WHEN YOUR JOB IS NO LONGER A "THING"

Of course, none of this may be comforting to you personally. In an episode of *30 Rock*, Liz Lemon, worried that her skills as a television writer aren't wanted anymore, wanders the streets, trying to convince random strangers that the written word matters and they really should be willing to pay for it. Suddenly, in a fantasy sequence, she's confronted by a vision of other people "whose professions are no longer a 'thing.'" That includes a travel agent, an American autoworker, and a guy who used to "play dynamite saxophone solos in rock-and-roll songs." "Come with us," the travel agent says. "We live under the subways with the CEO of Friendster."[8]

If it's your job that's suddenly out of date, what do you care if somebody else gets a job in a new industry? You only care if *you* can get a job in a new industry. Even in the story of John Henry, the steam hammer was undoubtedly bad for the workers laying rail—but somebody, somewhere, was hiring people to build steam hammers. The problem was that it wasn't going to be John Henry who got the job.

It may be easy enough for certain people in publishing and broadcasting to move to Web production, but some of the other changes are harder to deal with. For example, the *Occupational Outlook Handbook* projects employment in utilities will fall 11 percent by 2018, and by 14 percent in mining, quarrying, and oil and gas extraction—in both cases at least partly because of improved technology. By contrast, employment in professional, scientific, and technical services is supposed to grow by 34 percent, with 2.7 million new jobs.[9] But how many people in mining or utilities are ready to move over to manage computer networks?

DON LOCKWOOD DOES FINE

Sometimes even people who theoretically *could* make the shift have a hard time with it. Remember *Singin' in the Rain*? Don Lockwood (the Gene Kelly character) made the transition from silent movies to the talkies just fine—he had a great voice. But the glory days were over for the nasal-voiced Lina Lamont. The movie actually has some basis in fact: there were a few widely admired silent screen stars whose careers took a nosedive when technology brought sound to motion pictures.

The good news is that most of us will get another job. People adapt, and so does the economy. Career change is difficult and time-consuming but possible. "There is no reason to think that technology creates unemployment," says

David H. Autor, an MIT economist and one of the leading experts on technology and the economy. "Over the long run we find things for people to do. The harder question is, does changing technology always lead to better jobs? The answer is no."[10]

That leads to the two bigger questions here, especially in terms of the ways this current round of technological change may be different from previous ones. First, what kind of jobs does technology create, and what kind does it destroy? And second, are the new technologies coming online inherently creating fewer jobs than the ones in the past?

STUCK IN THE MIDDLE WITH YOU

The working assumption, for most experts in the field, is that technology favors the smart, the creative, and the well-educated. The management guru Peter Drucker made this case in the 1950s, when the first computers filled an entire room and could barely do what a throwaway pocket calculator can do now. "They will not make human labor superfluous," Drucker wrote. "On the contrary, they will require tremendous numbers of highly skilled and highly trained . . . managers to think through and plan, highly trained technicians and workers to design the new tools, to produce them, to maintain them, to direct them."[11]

Put another way, this is about the difference between what machines do well and what people do well. Autor and his coauthors contend that it's about the difference between routine, repetitive tasks and tasks that require problem solving and communication. If you're doing routine, repetitive work, such as working on an assembly line, picking vegetables, or handling deposits or withdrawals at a bank counter, a computer or another machine might do your job much better. If you're in charge of making sure those tasks get done, or writing about them, or marketing them to others, a computer may help you do your job, but it can't do it instead of you.[12]

If you think about it like that, you get this kind of chart:

Impact of Computers on Tasks

	Routine tasks	Non-routine tasks
Analytic and interactive tasks		
Examples:	•Record-keeping •Calculation •Repetitive customer service (e.g. bank teller)	•Forming/testing hypotheses •Medical diagnosis •Legal writing
Computer Impact:	Substantial substitution	Strong complementary
Manual tasks		
	•Picking or sorting •Repetitive assembly	•Janitorial services •Truck driving
Computer Impact:	Substantial substitution	Limited opportunity for substitution or complementarities

SOURCE: MIT Press Journals, "The Skill Content of Recent Technological Change: An Empirical Exploration," November 2003.

Whether a machine can do your job depends on what the job is—and how routine and repetitive it is.

As you can see, some very high-skilled jobs would seem pretty safe. So do some low-skilled jobs. But a lot of jobs in the middle are in trouble.

And if this looks a lot like the ways globalization affects jobs, as we discussed in Chapter 12, that's no accident. Computer technology—since it means certain jobs can be done anywhere—is a major factor driving globalization. You've probably seen it in your own life. The automated teller machine has dramatically changed how we withdraw and deposit money from the bank, which is pretty much a matter of record keeping. That was bad for tellers, but not necessarily for other people in the banking world, such as loan officers. To get a mortgage, which should require a judgment about whether someone's worthy of credit, usually requires talking to a person—although given the "liar's loans" and shoddy practices that led up to the financial meltdown in 2008, computers could hardly have done worse than the loan officers. On the other end of the scale, banks employ just as many janitors as they always did, because you can't outsource or automate cleaning the office every day (with all due respect to the Roomba).

This also underscores the value of a college education, as we talked about in Chapter 9. MIT economists who've been examining changes in the workplace say that technological change explains 60 percent of the shift toward college-level work that's been documented since 1970.[13] At least so far, computers have been primarily used for analytical jobs (see table on page 228), and because analytical jobs usually require a college education anyway, that makes perfect sense.

Finally, there's a much more positive word for all of this: *productivity*. Technology enables business to do more work with fewer people, which means at less cost and with greater efficiency. Economists love productivity; it's one of the best signs of a healthy, vibrant economy, and over time, it's the engine that creates a rising standard of living.

THAT HOLLOW FEELING

In fact, that's what lots of businesses have been doing over the past few troubled years: trying to become more efficient by shedding workers and embracing technology. That's also one reason why employment bounced back slowly after the Great Recession, argues Mark Thoma, an economist at the University of Oregon. When businesses use technology to become more efficient in bad times, it means they don't have to bring as many workers back when times improve.[14]

In fact, experts such as MIT economist Tyler Cowen say one of the big trends of the next few decades will be fewer people doing more work. A lot of the jobs eliminated during the recession were in positions where technology can actually do the job better—so why would businesses bring those people back? "It's not a question of getting back to where we were, but rather that the economy must solve a new problem of re-employing a lot of people who were not, in reality, producing very much in the first place," he writes.[15]

Another way of looking at this is whether technology is actually starting to redefine what's routine and what's not. On one hand, computers are encroaching on the lower-end retail jobs that used to require personal contact—or the retail outlet is actually disappearing as more and more people buy products online. Technology is already pushing a lot of tasks that businesses used to have to hire people to do onto the customers themselves. Anytime you've gone through the self-checkout line at a store or checked in and printed out your own boarding pass at home or on a machine at the airport, you've seen this at work.

On the other hand, computers may also be moving into the analytical work as well. IBM's Watson computer is designed to break one of the big barriers between the computers in science fiction and the one that's sitting on your desk right now: it can work with natural language. In other words, you can ask it a verbal question and it can give you an answer.

With a skillful eye toward promotion, IBM got Watson scheduled as a contestant on the long-running game show *Jeopardy*, where it defeated two of the show's top human champions. "I, for one, welcome our new computer overlords," former champion Ken Jennings said dryly. Watson wasn't perfect—it missed some answers, and some suggested the real secret to its success was that as a machine, it could push the answer button much quicker than the humans.[16]

DR. WATSON SANS SHERLOCK

But IBM isn't betting Watson's profitability on winning a lot of trips to Cancun or a lifetime supply of Turtle Wax on game shows. The company argues that one of the great uses of Watson will be in health care, where it can take medical histories from patients and even make basic diagnoses—jobs now performed by nurses, physicians' assistants, and even doctors. The idea of Watson becoming Dr. Watson may unnerve some patients, but it could turn out to be extremely helpful given the relentless escalation in the nation's health care costs. And that would have implications in other fields, such as teaching.[17] Some analysts argue that technology could go even further in health care. Computers are becoming more and more capable at visual pattern recognition, or looking at images and discerning shapes. That's essentially what a radiologist is doing when she's looking at an X-ray or an MRI result. It could be another kind of analytical job that potentially could be done by technology.[18]

Lawyers will be affected as well. (Stop that cheering.) "E-discovery" software can run through evidence such as legal documents and e-mails and look for relevant patterns—a job that used to be delegated to paralegals and junior associate lawyers. Some computer experts have even designed e-discovery software to search for potential fraud. Using the enormous e-mail trail left behind by the Enron scandal as a base (the so-called Enron Corpus, which is fun to say out loud), software can now look for shifts in tone, phrasing, and even the kind of communication ("Let's talk about this offline") that might indicate something sleazy is going on.[19] And of course, college professors are already using software to ferret out student plagiarists. This not only saves professors time but also puts the fear of God into potential cheaters, which is entirely a good use of technology, in our opinion.

All this goes to what economists call "hollowing out," where both high-wage, high-skill jobs and low-wage, low-skill jobs are growing, but the jobs in the middle suffer. That has serious implications by itself, but what if the middle that is being hollowed out is becoming increasingly larger? As technology performs more tasks, what happens to the people who used to do those jobs? Do some people in these "middle jobs" slide down the economic ladder, with file clerks and bookkeepers forced to take lower-wage jobs in manual labor and retail? Does that force the wages for lower-skill jobs even lower and make it harder for people to make ends meet?[20]

And what about higher-skill jobs? If you've gone to school for years to become a radiologist, you can't just switch careers overnight, even to another medical field.

The economist and *New York Times* columnist Paul Krugman says this expansion of technology into higher-skill jobs calls into question one of the basic assumptions we've been making about the job market: that getting more education and more skills makes you more marketable. "The idea that modern technology eliminates only menial jobs, that well-educated workers are clear winners, may dominate popular discussion, but it's actually decades out of date," Krugman says.[21]

WHERE'S THE NEXT BIG THING?

You may think technology is an unstoppable force reshaping our lives, but the question some economists have been raising is whether the era of big change is over, which might mean the era of big booms in employment is also over. In his book *The Great Stagnation*, Cowen argues that in the first half of the twentieth century, the world came up with huge advances that made enormous differences in how people lived, such as the automobile and electricity. Since then, we've mostly been building on that legacy.

> Today, in contrast, apart from the seemingly magical Internet, life in broad material terms isn't so different from what it was in 1953," he wrote. "It would make my life a lot better to have a teleportation machine. It makes my life only slightly better to have a larger refrigerator that makes ice in cubed or crushed form. We all understand that on a personal point of view, yet somehow we are reluctant to apply it to the economy writ large. But that's the truth behind our crisis today—the low-hanging fruit has mostly been plucked, at least for the time being.[22]

Cowen's point is that the Internet, and all the products it has brought with it, is dramatically changing people's lives, but not necessarily creating jobs. Internet technology inherently allows businesses to serve more people with fewer employees, since it automates so many tasks and pushes them off onto customers. Facebook, for example, is able to serve more than 300 million members with only 1,700 employees. Twitter, with about 9 million users, has only 300 employees. Compare that with older communications technologies such as the telephone, which required tens of thousands of workers to serve hundreds of millions of customers.

Critics argue that Cowen underestimates all the jobs Facebook and other social media technologies create indirectly: the Web designers, app developers, social media marketing specialists, and others who are all trying to make a buck using these platforms. It's also famously difficult to predict what will come next in technology. As Cowen rightly points out, despite the best predictions of the 1950s (or *The Jetsons*), we're not living on the moon or flying around with jet packs. Then again, when the telephone was first invented, no one could have predicted that 130 years later people would get paid to create games such as Angry Birds to play on a phone you carry in your pocket.

RIDING IT OUT

Barring some sort of *Mad Max*–type apocalypse, technological change isn't going to stop. Maybe it'll speed up, maybe it'll slow down, but the fact of its existence has been a central part of our lives for more than 150 years now. We're not going to be Luddites, nor are we likely to be living in the Matrix. But there's no reason to assume that the information revolution is going to be any less disruptive than the industrial revolution, which upended much of Western society. We know, pretty certainly, that any job that you can imagine a machine doing may be done by one—someday.

But we also can't imagine the new jobs technology will create either. In 1850, nobody knew about moving pictures, so they couldn't even imagine a projectionist—or a cinematographer, a sound engineer, or a key grip. Even twenty years ago, there was no such thing as a social media coordinator or an app developer.

What we don't know is which jobs, specifically, will be born or die, nor do we know how to help the people whose jobs tend to get lost along the way.

Going, Going, Gone

Paul Revere was a silversmith, we know that, but honestly, when's the last time you met one? Technology destroys jobs all the time and then creates new ones as progress marches forward. Here are a few jobs that were once pretty common and are now pretty rare, along with some that may be gone in just a few years.

"Copy!"

Sometimes jobs just change. Copy boys, including this rather elderly example, used to work on every newspaper in America, carrying paper copy, Telex messages, and photos from reporter to editor to composing room. There's less paper, but editorial assistants still exist for the grunt work in publishing, and arguably are treated worse. Watch *The Devil Wears Prada* sometime.

The Ice Man Cometh

Once people kept food cold with iceboxes: big coolers in the kitchen, powered by huge blocks of ice. Icemen (including the grandfather of one of your authors) delivered blocks of ice to homes and restaurants. But as electric refrigerators came in during the 1940s and 1950s, the ice routes died out.

"Number, Please"

For the first forty years or so of the telephone era, you had to ask a switchboard operator to place your call for you. The first phone operators were hired in 1878, and even after direct dialing was introduced in the 1920s, you still needed operators to reach extensions in big buildings, or to make long-distance or collect calls. Now operators are like bank tellers: they're still around, but when was the last time you talked to one?

The Smithy

It used to be that every town, even a small one, had a blacksmith, because otherwise there was no way of fixing anything made of metal. Granted, people still repair metal (next time you're in a fender-bender, think about what's happening at the auto body shop). But almost all of this work is done at factories now.

All photographs: Library of Congress.

||||||||||||||||||||||||||||||||||||||||||||||||||||||||||||||||||||||

CHAPTER 14

WILD CARD #3

Would Reducing Immigration Reduce Unemployment?

A simple way to take measure of a country is to look
at how many want in . . . and how many want out.

—*Tony Blair*

If you look back at the history of immigration in the United
States, one of the most striking things is that we've *never*
been able to make up our minds about whether it's a good
thing. Just think about the famous 1886 inscription on the
base of the Statue of Liberty that nearly every American
learned in school: "Give me your tired, your poor, your
huddled masses yearning to breathe free." It's one of the
most heartwarming expressions of our romantic notions of
immigration—and yet a mere four years earlier Congress
enacted the first immigration restrictions, including a near-
total ban on immigration from China.

But you don't have to go back to the nineteenth century to see this ambivalence at work. You just have to read
the news about Congress's repeated, failed attempts to pass
immigration reform, controversial laws in Arizona and

Alabama to clamp down on illegal immigration, and the so-called minutemen patrolling the border with Mexico. About the only bipartisan immigration effort over the past decade has been the effort to build some seven hundred miles of fence to secure a two-thousand-mile border.

Obviously, there are a lot of issues wrapped up in the immigration debate, but in this book, we're only worried about one specific angle: what does immigration mean for jobs? Do immigrants, legal or illegal, take jobs away from native-born Americans, or drive down wages because they're willing to work for less? Do immigrants actually create jobs by starting new businesses and bringing new ideas to our economy?

Most economists would say that immigration is a boon to the U.S. economy overall. But, as you've already seen if you've gotten this far in the book, whether immigration is good for you personally is another matter.

WHERE WE ARE NOW

The United States is in the midst of one of the greatest waves of immigration in its history—not quite as big as the wave that arrived in the late nineteenth and early twentieth centuries, but huge by any standard. Consider this:

- We've been admitting about 1.1 million "legal permanent residents" a year, according to federal statistics.[1] Roughly 12.5 percent of the 307 million people in the United States are now foreign-born, the highest proportion since the 1930s.[2]
- Immigrants are an even bigger part of the working population; in 2009, about one in seven members of the workforce was foreign-born.[3]
- The most common reason people immigrate to the United

States is because they have family members here, particularly spouses, parents, or children. That accounted for two-thirds of all new legal permanent residents in 2009.[4]

Immigration

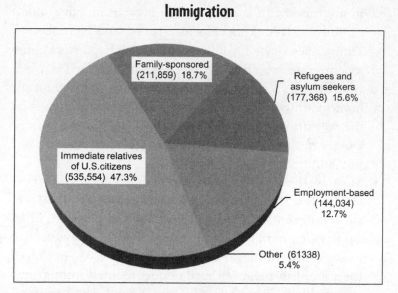

SOURCE: Department of Homeland Security.

Only a relatively small portion of legal immigrants are admitted on employment grounds. But of course, the other immigrants get jobs once they're here.

• Only 14.2 percent of legal permanent residents were admitted for employment reasons, including "priority workers; professionals with advanced degrees or aliens of exceptional ability; skilled workers, professionals (without advanced degrees), and needed unskilled workers; certain special immigrants (e.g., ministers, religious workers, and employees of the U.S. government abroad); and employment creation immigrants or 'investors.'" All told, employment entries are capped at 140,000 people a year. Of course, that doesn't mean that immigrants who enter the country because of family connections or other legal reasons don't work. Obviously, many of them do.

- Immigrants mostly live in a relatively small number of "gateway states." In 2009, for example, 58 percent of all immigrants with legal residence lived in just five states: California, New York, Florida, Texas, and New Jersey.[5] That makes sense if you think about it: if you were going to emigrate, you'd probably gravitate toward a place where there's a support system of other people from your home country. That also means immigration's impact on the workforce varies widely as well. In California, one worker in three is foreign-born; in West Virginia, it's one in a hundred.[6]

- Most of the immigrants in the United States are here legally. But there are also an estimated 11.2 million illegal immigrants in the United States.[7] Of course, it's illegal to hire an undocumented worker; that's why you had to show proof of legal status the last time you got hired for a job. But most of those undocumented immigrants are working, either by using fake Social Security numbers or in "under-the-table" work for cash.

WHAT HAPPENS WHEN IMMIGRANTS COME TO TOWN

People who worry about the economics of immigration largely focus on two potential problems. One is that immigrants compete with native-born Americans for the limited number of jobs available and raise unemployment rates overall. The other is that even if immigrants don't take jobs away from people born here, they drive down wages for everybody, including themselves, because they expand the overall pool of labor and, possibly, are willing to work for less.

The extremely oversimplified theory would work like this: Say there's a city with 100,000 potential workers and

95,000 jobs. That would add up to 5 percent unemployment. Now suppose this city had 5,000 immigrants arrive. Either unemployment would double, to 10 percent, or employers would create more jobs, but be able to pay their workers less because there would be more workers applying for the jobs available.

Part of the impact, however, depends on who those immigrants are and what skills they bring. If, for example, the new immigrants are bakers or auto mechanics, that's going to mean more competition for the bakers and mechanics already in town. But if you're an accountant or a hospital technician, an influx of bakers wouldn't change your job prospects.

Except that labor markets actually don't work this way at all. And the reason we know this is thanks to, of all people, one of the world's last and loudest communists, Fidel Castro.

In 1980, the Cuban dictator decided that the best way of dealing with people he didn't like or who didn't like him was to let them leave. That became the famous Mariel boatlift, where 125,000 Cubans fled the ninety miles to Florida on whatever would float. Under U.S. law, refugees from Cuba get asylum automatically if they reach American soil. Some accused Castro of taking the opportunity to empty his jails, but most "Marielitos" were clearly just ordinary people who wanted to get out of Cuba. About half of them settled in Miami, swelling the local workforce by 7 percent.[8]

By the traditional theory, this should have increased unemployment or lowered wages in Miami. But it didn't. When one famous economic study compared what happened in Miami to other cities that hadn't seen an immigration surge, such as Tampa, Atlanta, Houston, and Los Angeles, the results showed a very different picture. Yes, unemployment among Cubans in Miami increased in 1981, after the Mariel refugees arrived, but it had gone back to normal by 1985. Wages for workers in Miami didn't change much.

What happened, the study concluded, was that the Miami labor market adapted and absorbed the new workers. Remember, the immigrants who come into a community aren't just workers, they're also *consumers*. They buy things, and that creates more demand, which in itself creates more opportunities for businesses to expand and create more jobs. In the short term, there can be trouble adjusting, but in the longer run, labor supply and demand in a specific area balance out. Studies examining similar situations in the United States and other countries (such as Israel, which took in hundreds of thousands of Russian Jews after the fall of the Soviet Union) have found similar results.[9]

WHO'S COMPETING WITH WHOM?

Other economists, however, point out that this "everybody adapts and wins" theory isn't the only possible explanation. A lot depends on who's competing for which jobs. For example, there's no question that Mexican immigrants, the largest single bloc of both legal and undocumented immigrants, generally have lower education levels than other immigrants and most native-born Americans. Some 41.5 percent of Mexican-born people living in the United States have less than a ninth-grade education, compared to 20.7 percent of the foreign-born overall and 3.6 percent of the native-born.[10]

For that reason, there's been a lot of focus among economists on what immigration means for lower-skill, less-educated Americans. There are studies that show immigration hurts low-skilled native-born workers, at least to a small degree. When the National Research Council looked at this in 1997, they concluded that immigration may have reduced the wages of competing workers by 1 to 2 percent.[11] Another study, however, concluded that a 10 percent increase in the number of workers in a specific skill group could reduce the weekly wage of competing workers by about 4 percent.[12]

A lot of the impact depends on how people respond when faced with new competition from immigrants. Maybe when a surge of immigrant labor moves into an area, the lower-skilled native-born workers already there move on to other towns with more opportunities. Maybe they go back to school to pick up more skills and change careers. Or it's possible that immigrants and unskilled native-born people aren't actually competing at all, because they're not doing the same jobs. A lot of experts argue that immigrants are "taking the jobs Americans don't want," and there's evidence for this. Less-educated native-born people tend to work in manufacturing or mining, while less-educated immigrants usually work in agriculture or personal services such as housekeeping, gardening, or food service.[13]

PROMOTIONS FOR THE NATIVE-BORN?

Or maybe the least-educated immigrants are really competing for low-skill jobs with the much larger pool of Americans who have high school diplomas, which actually puts the immigrants at a disadvantage. The native-born and higher-skilled immigrants, in this theory, end up as the foremen and shift managers for the new immigrants, because they've got the language skills and experience on the job. At least one study shows that in states with high immigrant populations the native-born end up moving into these "communications" tasks over time.[14] If that happens, over the long run immigration would actually drive up wages and create opportunities for native-born people. The real competition would be between new immigrants and the ones who are already here, and the challenge would be for immigrants to improve their skills to stay ahead of those who come after them.

There's also the separate issue of what would happen to high-skill workers if we encouraged more immigration. Why not bring in more doctors, scientists, engineers? There's always room for those folks, says the conventional wisdom. There's even a special immigration visa, the H1-B, that's designed to bring in people with technical skills, and business leaders seem to be testifying before Congress on a regular basis asking for more immigration in this area.[15] In 2008, there were more than 276,000 H1-B visas issued, nearly half of them for information technology professionals.[16]

The caution we'd raise is that there's actually less research on the impact of high-skill, educated workers on native-born employment. The Congressional Research Service warns that, given the limited research, there might be unintended consequences to focusing on bringing in high-skill workers.[17]

In some areas we're already gone beyond opportunity to dependence on high-skill immigrants. American university graduate programs are full of foreign students. National Science Foundation statistics show that in 2006, foreign students on temporary resident visas earned 32 percent of the doctoral degrees in the sciences and nearly 60 percent of the doctoral degrees in engineering.[18] We could certainly do more to channel American kids into science and math careers—but for the foreseeable future, we're going to be getting a lot of our technical talent from immigrants.

WALLING OUT INNOVATION

There's another angle to immigration, however: Does immigration actually create jobs? If so, what might we lose if we pull back? Remember, as we discussed before, immigrants aren't just job seekers. They're consumers, producers, and innovators. Consider this:

- The National Research Council concluded in 1997 that immigrants added as much as $10 billion to the economy every year.[19]
- A Duke University study found foreign nationals residing in the United States were listed as inventors or co-inventors on 24.2 percent of all U.S. patent applications in 2006.[20] Another study found that a 1 percent increase in immigrant college graduates resulted in a 15 percent increase in patents.[21]
- Between 1995 and 2005, immigrants founded one in four engineering and technology firms, generating an estimated $52 billion in revenues and employing 450,000 people.[22]
- Some 26 percent of U.S. Nobel Prize winners between 1990 and 2000 were immigrants, as were 25 percent of the founders of publicly owned American companies backed by venture capital between 1990 and 2005.[23]

The list of immigrants who've made big contributions to the United States in business and science gets very long indeed: Albert Einstein, Andrew Carnegie, I. M. Pei, Madeleine Albright, Isaac Stern, and on and on. And some have started enterprises that create a lot of jobs: Yahoo was cofounded by a Taiwanese American, Jerry Yang; Russian American Sergey Brin helped create Google; and eBay was launched by Pierre Omidyar, who was born in France.[24]

One of the secrets to our success as a nation has been in allowing people to come here and bring their ideas with them.

That "invite everyone, try anything" spirit has proved extremely useful, and some experts argue that the real danger about immigration and jobs is that the United States isn't as appealing a destination as it used to be. As countries such as China and India develop their own economies, there are more reasons for smart entrepreneurial people to stay at home and build businesses there, instead of coming to the United States.

Recently in the *Atlantic*, author James Fallows, who has spent extensive time in China, examined claims of American decline in "How America Can Rise Again." His conclusion: "The American advantage . . . depends on two specific policies that, in my view, are the absolute pillars of American strength: continued openness to immigration, and a continued concentration of universities that people around the world want to attend." Even if foreign students come here and end up going home again, they'll still be taking American ways of thought with them, and will want to build businesses that work with us, not against us, Fallows wrote. "We scream about our problems, but as long as we have the immigrants, and the universities, we'll be fine," one American in China told Fallows.[25]

ONE IN SEVEN MEMBERS OF THE WORKFORCE

As we pointed out when we started this chapter, there are a lot of different reasons people are worried about immigration to the United States, and only some of them are economic. Many, perhaps, most Americans, are uncomfortable with an immigration policy that seems to be routinely ignored and unenforced. On the jobs front alone, however, most economists argue that immigration is an overall plus to the U.S. economy. The debate is over the impact on specific groups, whether low-skill or high-skill workers.

More important, most of the research suggests that the negative impact of immigration on the job market is relatively small, while the potential payoff from new ideas and businesses started by immigrants could be huge. And you never know where the idea for the next Google or eBay might come from. Practically speaking, immigrants are already a huge part of our economy: one in seven members of our workforce. In fact, many major corporations plan advertising and marketing campaigns specifically designed to appeal to immigrant Americans or the U.S.-born children of immigrants. The Association of Hispanic Advertising Agencies, for example, projects that spending by Hispanic Americans will top $1.5 trillion in 2015—about 40 percent of whom are born outside the United States.[26] It's a lot of people and a lot of potential business.

Clearly immigrants create jobs, and we create jobs for immigrants. Those one in every seven jobs didn't come from thin air, and we didn't fire one in every seven native-born workers to do it. To the extent we're already expanding our economy and creating jobs as a nation, immigrants are helping us make it happen. And unless we're willing to forgo all the benefits and advantages of being one of the most popular immigration destinations on the planet, the real question is how we build on this and put it to our advantage.

CHAPTER 15

WILD CARD #4

The Aging of the Boomers and What It Means for Jobs

Ah, but I was so much older then,
I'm younger than that now.

—*Bob Dylan**

There is a man with a computer cable coming out of his head on the February 21, 2011, cover of *Time*.[1] The headline is pretty eye-catching too: "2045: The Year Man Becomes Immortal." The lead article discusses how advances in computing and nanotechnology could extend human life, or alternatively, how in the future we may be able to transfer our minds and thoughts into computers, so we don't have to worry about the old bod falling apart. It's *Star Trek* stuff, but longer life expectancies and better health as people get older have already changed American society—jobs included.

---

* Bob Dylan wrote the words. You probably recall the Byrds singing them. Their cover of Dylan's song inspired the title for their fourth album, *Younger Than Yesterday*, www.bobdylan.com/songs/my-back-pages.

It's not just that people are living longer and healthier. It's also that the superhuge baby boom—the roughly 78 million Americans born between 1946 and 1964—are the ones doing it.[2] Depending on which expert you listen to, either the boomers will be retiring in big numbers over the next two decades, thereby solving the country's unemployment problems, or they'll hang on to their jobs for as long as possible, making our problems worse. In this chapter, we cover some of what's being discussed among the experts who worry about demographics and the labor force.

A word to the wise as you work your way through this chapter: More than any other issue, there are major disagreements between the experts on this one, partly because there are so many unknowns, or, to use the famous Donald Rumsfeld formulation, so many "known unknowns." Experts can make educated guesses about what will happen, but even the smartest and most conscientious don't know exactly how people in this age group will respond to the dramatic changes we've been seeing in the economy. Our advice is to treat these forecasts the way you treat those insider predictions on the Academy Awards. People who assiduously track the pre-Oscars buzz are probably going to be right about the winners in most categories, but it's not a sure thing. There are always some unexpected wins and losses.

DEPENDING ON CIRCUMSTANCES

There are several moving parts here, and depending on how they move, the aging of the boomers could improve the country's unemployment picture, or it could cause even more problems. One issue is how quickly the boomers retire—whether they start leaving their jobs en masse once they hit retirement age or whether they decide to work longer. There is also the issue of whether the younger people in the workforce have the skills needed to fill the jobs that open up when the boomers leave. And, of course, there's the trend we've been examining throughout the book: whether employers will replace the retiring boomers, or whether at least some of their jobs will be eliminated through technology or offshoring.

Here are some of the competing predictions about what the aging of the boomers means for jobs and joblessness.

Prediction No. 1: As the boomers begin to retire in larger and larger numbers, unemployment rates will fall naturally.

The first merry band of boomers (those born in 1946) turned sixty-five in 2011, and over time, presumably more and more of them will leave their jobs and begin eating early-bird specials, babysitting the grandkids, taking up bridge or bingo or ballroom dancing—or skydiving, like the first President Bush. We won't delve into the massive debate about what this means for Social Security,* but as far as the workforce goes, some experts say that when the boomers start retiring in big numbers, we'll actually have a labor shortage.

Labor economist Barry Bluestone is one of them. He believes that if the American economy picks up steam and immigration remains at current levels, there will be "more jobs than people to fill them" as early as 2018.[3] If you accept Bluestone's assessment, there could be as many as five million positions that open up because of boomer retirements,[4] and other analysts see the same dynamic. Drew

---

* But we have done so elsewhere. Check out *Where Does the Money Go? Your Guided Tour to the Federal Budget Crisis* (New York: HarperCollins, 2011) for more than you'd ever want to know about this question.

Matus, an economist for USB Securities, predicts that the nation's uncomfortably high unemployment rates will come down faster than expected as boomer retirements get into full swing.[5] Matus believes a lot of economists are missing what will be a major, long-term structural shift in the labor force, one that will lessen unemployment.

So can we break out the champagne and stop worrying about how to prod the economy to create jobs? If more people retire, there should be more job openings, right? What's more, the next generation of workers—the so-called baby bust group—is smaller than the baby boom. It has about forty-five million people compared to the more than seventy million boomers, so there should be fewer people competing for the jobs available in the economy.[6] Could this be our salvation?

The pros that follow these demographic factors aren't celebrating yet. For one thing, although a labor shortage sounds like it could be a good thing—there would be lots of job openings, and employers would have to offer good wages to compete for the workers available—there are potential downsides. If employers have to raise wages considerably just to get warm bodies on the job, they would no doubt pass that cost along to customers, making goods and services more expensive. This could also make U.S.-produced products less competitive internationally.

But once again, economists don't necessarily agree on the degree to which a labor shortage of any magnitude is likely (see Prediction No. 2) or whether the boomers will stay in the workforce longer than earlier generations (see Prediction No. 3).

Barry Bluestone argues that "a labor shortage of this magnitude . . . could have a significant impact on our economy and on the quality of life in our communities,"[7] especially since the shortages would be pronounced in some service fields, such as teaching, nursing, and the clergy.[8] Bluestone is worried enough about the shortfall to argue that we ought to encourage these older workers (sorry, boomers—we know it's a shock to see yourself referred to as an "older worker") to stay in their jobs. Beyond that, he argues, we should look for ways to persuade boomers who retire from other fields to enter these service sectors as "encore" careers.[9]

Another wrinkle is that while the "baby bust" generation that followed the boomers is smaller, the "echo boom" following them is almost as large. This group of some 72 million people—basically the children of the boomers—began entering the workforce in the 1990s. What that means, according to the Congressional Research Service, is that the mismatch may not be as great as it initially appears.[10] Plus, with those other worrisome projections about the number of jobs going offshore and being eliminated by technology, it's too soon to relax. As we said, there are a lot of factors here, and even the major wonk types have a hard time predicting exactly how everything will shake out.

Prediction No. 2: There won't be an overall labor shortage, but there could be acute problems in some fields.

Boomers make up about a quarter of the U.S. population,[11] and if they all retire when they turn sixty-five, the process will last from 2011 through 2029.[12] It's just a big bunch of people who have spent years in the workforce and who will be leaving it over two decades. If you made a graph of it, it might look something like the Little Prince's drawing of the boa constrictor that swallowed the elephant. (If you don't remember it or haven't been reading the book to your children or grandkids lately, take a look the next time you're in a bookstore.)

Before the Great Recession, when unemployment

was historically low, the Congressional Research Service investigated whether the aging of the boomers would leave employers scrambling to find workers to replace them. Their conclusion was that the possibility of a broad, disruptive labor shortage was fairly low.[13] They point out that although the boomers will be retiring, the demographics will even out as soon as the children of the boomers start working in large numbers. CRS also points out that the United States has something of an ace in the hole here: there are millions of people worldwide who would love to come here and work. If we get into a bind, we can easily invite more immigrants to join us.[14]

The more worrisome problem, according to the CRS, is that there could be spot shortages, or at least tough transition issues, in some fields. The CRS focused on "mature industries" that have been around awhile and consequently have lots of boomers working for them. Emerging industries (the CRS used biotechnology as an example)[15] tend to have younger workers because they haven't been hiring people for as long a period of time.

So what fields will be seeing a lot of boomers head out the door? According to the CRS, nearly 60 percent of employees working in utilities are boomers, including 73 percent of the managers and 60 percent of the people who do installation, maintenance, and repair.[16] More than half of the people working in public administration are boomers, including 64 percent of managers, 56 percent of business and financial workers, and 69 percent of those who do government construction work.[17] Some fields—nursing is a prime example—have already been coping with labor shortages, and health care leaders are bracing themselves for even bigger shortfalls when boomer nurses start leaving the field in large numbers. Unless, of course, they really get that IBM Dr. Watson computer technology up and running extremely quickly.

In nursing, the problem is actually twofold. There has already been a growing need for registered nurses. According to 2009 estimates from the Bureau of Labor Statistics, the country will need more than half a million new RNs by 2019.[18] In fact, with the U.S. population aging, the need for health care workers of all kinds is projected to go up. According to the BLS, ten of the twenty fastest-growing fields of employment are in health care.[19] As of 2012, nearly a quarter of current RNs are in their fifties, and presumably they'll be retiring as they get older.[20] As boomer nurses retire, and as the country's aging population requires more health care, it's a double whammy to the system.

There are of course ways to address these spot shortages. Offering better salaries and working conditions would attract more people into the field. But the challenge is tougher when you're dealing with professions that require significant training. You can't just take anyone who decides that he or she would like to be a nurse and throw that person into the hospital. A prospective nurse needs to complete a nursing degree, and for more advanced positions, that means at least four years for an RN degree and in some cases graduate study as well.[21] People hoping to become nurses need to be able to pay for their education. You also need enough nursing education programs to accommodate them. In 2010, nursing schools turned away more than sixty-seven thousand qualified applicants because they didn't have enough qualified nurses to teach the next generation.[22]

Prediction No. 3: The boomers won't retire at the same pace as earlier generations. They'll try to hang on to their jobs.

The biggest "known unknown," of course, is whether the boomers will actually retire as expected. There are two questions. One is whether boomers just look at life differently than their parents did and aren't as eager to retire as their parents were—some may want to continue working at least part-time.

The other is whether they can afford to retire at sixty-five given all the upheavals in the economy lately. Americans across the board lost jobs, homes, and savings in 2008 and 2009, and boomers, who are closer to retirement, have less time to recover. Although the stock market recovered in 2010, 401(k) plans lost about a third of their value during the 2008 stock market plummet. Some boomers, in an effort to protect their retirement savings, pulled their money out of stocks and consequently didn't benefit from the bull market of 2010.[23] One survey of boomers conducted by Boston College researchers found almost two-thirds reporting having less money for retirement after the stock crash than before.[24]

So can boomers really afford retirement? The Employee Benefit Research Institute (EBRI) conducts an annual retirement confidence survey assessing people's plans and preparations for retirement. According to EBRI's 2010 report, only 23 percent of workers over fifty-five have more than $250,000 in savings; just 14 percent of workers between forty-five and fifty-four have that much.[25] Most are worried. Just 13 percent of workers over forty-five say they have enough money to "live comfortably throughout their retirement years."[26] Even more disturbing, fewer than a third of those over fifty-five say they're "very confident" that they'll have enough money to take care of basic expenses.[27]

Faced with the prospect of living just on Social Security in retirement, many boomers may in fact decide to work longer and/or keep working even after they've started collecting Social Security. In the Boston College survey, about 40 percent of boomer workers said they planned on working longer because of the economic upheavals of 2008 and 2009. Most said they planned to delay retirement by at least four years.[28] Past surveys suggest that people's predictions about when they'll retire often change,[29] but clearly a number of boomers are contemplating working longer because of economic factors.

There's another reason boomers may be more uncertain. Compared to the previous generation, boomers (especially younger ones) are less likely to have pensions and more likely to have 401(k)-type retirement plans that can go up and down in value depending on how the stock market is doing.[30] Experts are divided on whether this trend away from pensions really makes workers less financially secure,[31] but many boomers may not *feel* as secure if they can't count on a guaranteed pension check every month. It's another fact that could lead boomers to stay in their jobs longer so they can beef up their 401(k) savings while they're still physically able to work.

And what do boomers really want? Would they retire right on schedule if they could, or do they actually want to work longer? After all, with longer life spans, many could be looking at retirements lasting two or three decades. Surveys show that 70 percent of workers, young and older, expect to work for pay after they retire, although many may not be planning on working full-time or at the same job.[32] Maybe the Pew Research Center, which has been tracking attitudes and expectations on retirement for years, has the best advice: "Public attitudes, expectations and experiences [about retirement] are in a period of transition . . . and this evolution in attitudes is likely to continue for years to come."[33]

Prediction No. 4: Employers will stand by willingly as their older employees leave the workforce. Or employers will try to hold on to

older workers and make a variety of changes to keep them on the job.

Whether boomers continue working past the traditional retirement age depends partly on whether they are encouraged to stay at work or are nudged along. It also depends in some degree on whether they can find jobs if they want them.

Right now at least, the statistics suggest that more people over sixty-five are in fact working than before the Great Recession. In 2009, there were 6.6 million people of retirement age in the workforce compared to just over 4 million in 2001.[34] And as we noted earlier in Chapter 3, the unemployment rate for older workers over fifty-five is actually lower than for younger workers. On the other hand, older workers who lost jobs in the recession have generally spent more time unemployed before finding a new job.[35] Experts at the Bureau of Labor Statistics caution that at this point, it is not clear whether these patterns are blips due to the poor economy or the first signs of a longer-term trend.

It's also not clear whether older workers have more difficulty finding new jobs because their skills are out of date, because they can't find jobs that match their experience and salary expectations, or because employers discriminate against them. The Equal Employment Opportunity Commission reported a 29 percent jump in the number of age discrimination complaints between 2007 and 2008.[36]

Sometimes the controversies turn on subtle judgments about whether older employees lack needed skills that younger workers possess (computer skills, for example) or whether companies just want to move older workers out so they can replace them with younger, less experienced, lower-paid employees. Disputes have emerged at some of the nation's more prestigious law firms. The firms claim that some of the older partners have lost their "edge." The older lawyers who were targeted believe they are being pushed out for financial reasons.[37]

The United States is hardly alone in having lots of older workers at or nearing the age of retirement. It's a worldwide phenomenon that is actually more pronounced in Europe and Japan. The Economist, for one, suggests that traditional models in which people get higher salaries based on how long they've been on the job are hurting both employers and employees, Instead, they suggest that pay-for-performance models—where workers are paid based on how well they perform rather than their years of experience—are preferable: "Companies are still stuck with an antiquated model . . . [that] assumes that people should get pay rises and promotions on the basis of age and then disappear when they reach retirement. They have dealt with the burdens of this model by periodically 'downsizing' older workers or encouraging them to take early retirement."[38]

But there does seem to be a countervailing movement designed to keep older workers working, although perhaps not in their traditional jobs or on their traditional schedules. The Economist points out that companies worldwide are beginning to realize that they could face a debilitating loss of skills when the baby-boomers retire en masse."[39]

Jeffrey Joerres, who heads Manpower, Inc., one of the world's largest employment consultants, sees the same challenge. He argues that employers will essentially be shooting themselves in the foot if they see boomer retirements as a way to replace costlier older workers with less expensive younger ones. "Employers still seem to view coming retirements as cost-saving opportunities," Joerres wrote in the Wall Street Journal in 2009. "This view is dangerous and shortsighted. Employers will need to shift their mindset and, in the short term, take steps to slow the exodus of older workers whose skills and knowledge are most valued."[40]

Some groups are taking the lead on figuring out how to keep the boomers thriving at work. The Conference Board, an influential business research and policy group, has launched a "mature worker" initiative designed to help "employers engage and develop mature employees within the rapidly changing multigenerational workplace."[41] Among the questions the Conference Board is studying are how to "recruit, engage, and retain mature workers" and "transfer critical knowledge from one generation to another." Civic Ventures, a nonprofit, is focused on luring retiring boomers into jobs "solving serious social problems—from education to the environment, health care to homelessness."[42]

Boomers are often credited, and sometimes resented, for the impact they've had on the country's culture and economy, just because there are so many of them. Being boomers ourselves, we have to say that we don't generally feel all that powerful. You could argue that boomers are far less influential than the tech-savvy younger generations who really do seem to be upending age-old ways of doing things, in the world of work and elsewhere. But we'll leave the boomers-versus-the-world battles for someone else to fight.

When the boomers leave their jobs and fully retire (whenever it takes place), their exit will no doubt have an impact. It's also true that at this point, no one knows for certain exactly what it will be.

CHAPTER 16

FOURTEEN BIG IDEAS FOR CREATING MORE AND BETTER JOBS

You can go with this,
Or you can go with that.
—*Fat Boy Slim, "Weapon of Choice"*

We've spent most of our time so far trying to offer some background information on what's causing the country's jobs problems and explaining why we face a long-term challenge—even if unemployment rates ease. We've also covered some of the thinking behind different ideas about solving the country's jobs problem, such as improving education or cutting red tape. Now it's time to get down to specifics.

In this chapter, we present fourteen proposals from across the political spectrum and summarize the pros and cons of each. Obviously, these aren't all the ideas out there (nor all the pros and cons), but we hope it's a fairly representative sample—one that will jump-start your own thinking and help you weigh the ideas the presidential and other candidates will be proposing in the 2012 election and beyond.

If you've been reading along carefully, you probably

have your own checklist of criteria for judging what's likely to help or hurt on jobs. But, just as a reminder, here are some considerations we think are vital:

- **Can we afford it?** All of these ideas have costs, both in dollars spent and in dollars not spent on other items. Many economists believe tax cuts of different kinds spur job creation, but that means less money coming into the U.S. Treasury and even larger federal deficits. Starting new programs in education or energy also adds to the red ink. On the other hand, despite everything, we are a very wealthy nation, and investing in job creation is probably one of the very best ways to use our money. Plus, a growing economy will help us address our budget problems because with prosperity, tax revenues increase—when people earn more, they pay higher taxes. So there are no easy, foolproof answers here; we're going to have to be thoughtful about what we can and can't afford.

- **How will it encourage employers to create jobs?** As we said up front, you can find experts who support all the ideas here, and you can find others who don't, so you'll have to use your own judgment. Just how convincing or direct is the connection between the idea that's being proposed and a person somewhere actually hiring someone? Because that's our aim here: having more employers add jobs.

- **How will the idea affect the country's ability to compete internationally?** This is the new reality we have to live with. Will the proposal put American companies in a better or worse position in the international markets, and how does that affect jobs here in the United States? This can be tricky because companies often need to slim down what they pay for labor so they'll be competitive, but having more people gainfully employed here drives up consumer spending, which is crucial for creating and holding on to jobs.

- **Will it create jobs that last?** As we point out in Chapter 4, jobs come and go all the time—that's part of a healthy economy. But there are important differences between ideas that create jobs to help the country get through a recession and ideas that create jobs that are likely to last for a good long while.

- **Will it create good-paying jobs so people earn enough to make ends meet?** It is certainly true that having a job, any job, is better than not having a job at all, but for most of us, the better goal is to create more jobs with salaries high enough so workers can live more stable and secure lives. This is an issue that often gets lost in surface debates over job creation, but it's an important consideration too.

- **This is not a "just pick one" list.** Chances are that we'll need to pursue a number of different strategies to begin to make a dent in our jobs situation. How do the ideas that you favor work together? Would some of them be more effective if they were combined—such as different ideas that aim to lower costs for businesses so they can expand and hire, or different ideas to spur innovation and growth? Try to come up with a package that seems sensible to you, keeping the uncertainties and limitations clearly in mind.

So let's take a look at some top contenders. We'll revisit some proposals we touched on earlier, so if you've actually read every single page we've written so far and have a memory like Spencer Reed in *Criminal Minds*, please forgive us. We are assuming most readers won't mind a little review. And if you don't see your favorite proposal here, we hope you'll visit the book website at www.wheredidthejobsgo.org

and post your suggestions there, alongside those of other readers. We expect there will be many more ideas for creating jobs emerging in the next few years (at least we hope so), so we'll be commenting there too.

Here goes:

1. KEEP TAXES LOW; RULE OUT ANY TAX INCREASES UNTIL THE UNEMPLOYMENT RATE DROPS BELOW 5 PERCENT

Experts who see low taxes as the number one prescription for job creation think it's time for the country to put its money where its mouth is, so to speak. They want Congress to lock in current tax rates until the jobs situation improves, and do it across the board—for higher-income people, lower-income people, employers, people who inherit money from their families, everyone. When wealthier people keep more of what they earn, they are more likely to invest in new enterprises (either directly or through the stock market). Plus they're big spenders, and that creates jobs. For the not-so-rich, lower taxes means people have more money left over after paying basic expenses, and that strengthens the country's consumer economy. Experts who support this approach believe that too many Americans are hanging on to their money (rather that spending it) and confining themselves to safe investments (rather than bold ones). They're also convinced that employers who aren't worried about higher taxes will be more likely to expand and hire and give their workers better pay and benefits. Remove the threat of higher taxes, they say, and you'll go a long way toward getting the economy humming and creating more jobs.

Sounds good, and low taxes are always popular, but critics make good points too. One is that we've had historically low taxes since the early 1990s, and it has done very little to spur job creation. Economist Michael Mandel points out that private sector employment rose by just 1.1 percent between 1999 and 2009, the lowest ten-year increase by far since the Great Depression.[1] Plus these historically low taxes certainly didn't save us from the worst recession since the Great Depression either. Even more important, critics say, keeping taxes at such historically low levels makes it almost impossible to address the country's huge federal deficits and spiraling debt. For this group, it's not a fear of higher taxes that is making investors, employers, and consumers cautious—it's the fear that the United States could face a debt crisis like the ones in Europe. Most experts in this camp dismiss the idea that raising income taxes on wealthier Americans will somehow stop employers from hiring.[2] They believe that reducing the deficit with a blend of tax hikes and spending cuts is what's really needed to strengthen the economy and pave the way for more jobs. And remember, 5 percent is roughly the natural rate of unemployment we mentioned earlier. Waiting for unemployment to drop below that to raise taxes may mean we'll be waiting a long, long time.

2. OFFER TAX BREAKS FOR COMPANIES THAT ADD JOBS

Rather than keeping taxes low across the board, some experts say we'd be much better off targeting tax breaks specifically to employers who hire people. There are two main variations to this idea. One is to give tax breaks to companies that are hiring by reducing the Social Security and Medicare taxes these employers pay for each worker. Currently, employers pay taxes of 7.65 percent on each worker's salary,[3] so the idea is that cutting or eliminating these taxes makes it much cheaper to add workers. When the Congressional Budget Office evaluated a number of different ideas for creating jobs in 2010, it predicted that this approach would be three times as effective as across-the-board tax cuts.[4] The other variation is to give employers specific tax breaks when they expand their workforces. The government had a program like this in the late 1970s that reduced the employer's costs for hiring new workers by about 20 to 25 percent.[5] The theory behind both is pretty much the same: most employers could probably use a few extra hands on deck, and if it costs less to hire new people, they will be more likely to do it.

There are two main drawbacks here. One is that it's generally seen as a short-term solution—something to perk up hiring until the economy is stronger. Doing this for a long period of time would be a huge new financial obligation. Letting employers off the hook for Social Security and Medicare taxes for even a short time chips away at the funding for those two critically important programs, and both already face serious financial issues down the line.[6]

The second trade-off is that the government (and the taxpayers) would basically be subsidizing hiring by private companies, and that creates a radical shift in the relationship between government and the private sector, raising all sorts of free market issues. What's more, giving tax credits to companies for hiring new workers turns out to be more complicated than it sounds. You could actually give less scrupulous employers an incentive to game the system by laying people off and then hiring new employees to get the tax credit. Trying to prevent this kind of abuse means these laws can be very complex. In the end, some critics say that tax credits won't make a difference if companies don't have plenty of customers and genuinely need the help—and in that case, they would probably be hiring new people anyway.[7]

3. LOWER BUSINESS COSTS BY KEEPING ENERGY PRICES LOW AND ROLLING BACK ENVIRONMENTAL REGULATIONS

Surveys show that Americans are split down the middle on whether protecting the environment or encouraging economic growth should be the bigger priority.[8] But some experts say that if we want American businesses to grow and create more jobs, we need to resist ambitious policies designed to curb global warming and other environmental problems. They argue that these policies will cost American businesses billions of dollars, making them less competitive internationally and less able to create jobs. For example, many environmentalists want to tax or regulate the use of traditional fossil fuels such as coal and natural gas (used to generate electricity) to encourage people to move to cleaner alternatives. But making common forms of energy more expensive makes the cost of doing business rise too. Plus there are the costs of complying with environmental regulations. One study put the average cost across all businesses, even without extensive regulation of fossil fuels, at about $1,500 per employee. That's more than double what businesses spend following health and safety regulations.[9] Not only do these costs add up, but this is an area where government can have a direct impact. As Rea Hederman and James Sherk, experts at the Heritage Foundation, point out: "While Congress can do relatively little about poor sales, it has direct control over both the tax and regulatory burdens businesses face."[10]

One trade-off here is obvious. Environmentalists argue passionately that regulation has dramatically improved health and quality of life for the American people,[11] and that it is urgent to reduce our use of fossil fuels such as coal, oil, and natural gas if we want to protect the planet for future generations. But there's a jobs argument here as well. If you want to create jobs, some experts say, one of the very best strategies is to move boldly into greener energy. Taxing or regulating fossil fuels not only helps the environment, these experts say, but allows all sorts of new alternative energy companies to gain a toehold. As they build their businesses, they'll be hiring. Green energy advocates also say that we're behind the curve internationally and need to catch up. Right now, companies in Europe, China, and elsewhere are way ahead of us when it comes to manufacturing solar panels and wind turbines and experimenting with ways to make using coal cleaner and nuclear energy safer. Our overreliance on traditional fossil fuels may offer some short-term cost savings, these experts argue, but in the long run there will be more jobs in developing new energy, not in clinging to the old.

4. BUILD THE ELECTRIC GRID

As we discussed in Chapter 10, many experts say one of the best ways to create jobs is for the country to improve its aging infrastructure. Obviously, there are a lot of other infrastructure needs in the country, but if you had to pick just one thing to work on, modernizing our creaky electric grid might be the one. The existing grid is already running close to capacity, and the Department of Energy has warned that "new lines need to be constructed to maintain the system's overall reliability."[12] What's more, it's a dinosaur—we need an advanced "smart grid" that uses modern computing to improve efficiency and performance. Building one, which would be paid for by federal, state, and local governments and the utilities, would provide jobs for a couple of decades. In 2009, the federal government allocated about $50 billion toward modernizing the grid, and according to the White House, even this relatively small investment will create "tens of thousands of jobs across the country," including "high paying career opportunities for smart meter manufacturing workers; engineering technicians, electricians and equipment installers; IT system designers and cyber security specialists; data entry clerks and database administrators; business and power system analysts; and others."[13] The gist of the argument is that we need a new grid and building one will provide jobs, so let's get started and let's do it right.

Again, one major problem is obvious. Fully modernizing the grid is estimated to cost between $1.5 and $2 trillion between now and 2030,[14] at a time when federal, state, and local governments are already reeling with debt. What's more, some experts say we shouldn't launch into building a new smart grid until we have a better understanding of how the new technology will work and what versions are the most efficient, reliable, and cost-effective.[15] The Center for American Progress (CAP), which is a strong advocate of building a new grid, points out another complication that even supporters need to keep in mind: we'll need lots of workers to build the grid, but we may not have enough people trained for this type of work. According to CAP, we'll also need "an increased and sustained commitment to job training and workforce development" to get this job done.[16]

5. GUARANTEE A MINIMUM WAGE

As of 2009, the minimum wage in the United States (which is established by Congress) was $7.25 an hour,[17] which means that a person working forty hours a week earns a little over $15,000 a year. As the nonprofit group Common Dreams points out, with pay at that level, a single mother with two children could work full-time and still earn less than the poverty level of $18,310.[18] For a lot of Americans that just seems wrong, and they say that raising the minimum wage would both help individual Americans—who are, after all, working and trying to support themselves— and spur economic growth.[19] There are roughly 4.4 million people working for the minimum wage or less.[20] The leisure and hospitality industry is most likely to employ minimum wage workers,[21] and the required pay rate is actually lower for waiters who earn tips ($2.13 per hour).[22] Advocates of raising the minimum wage say that putting more money in the pockets of these workers would help local businesses. These low-wage earners would be able to spend more, and that of course helps preserve and create jobs. Some also argue that raising the minimum wage is better for employers too, since paying employees better would reduce turnover and absenteeism, which can be costly for any business. There are a number of specific proposals on the table. During the 2008 presidential campaign, President Obama proposed raising the minimum wage to $9.50 an hour, which would basically bring the figure up to where it would have been if it had been rising with inflation over the years.[23]

There are several arguments against raising the minimum wage, and perhaps the most important one is that it reduces the number of job openings. Economists differ on this question, and to be fair, it's a lot harder than you might think to figure out whether having a minimum wage means fewer jobs. Ball State University economist Michael Hicks has looked at the impact of the most recent increases in the minimum wage, which went up by 14 percent in 2007, 12 percent in 2008, and 11 percent in 2009.[24] According to Hicks's calculations, these changes could have resulted in the loss of about 550,000 part-time jobs in 2008 and 2009. Hicks points out that until 2008, the legal minimum wage was actually lower than the prevailing wage for most entry-level jobs. The most recent adjustments, however, moved it higher than what employers were used to paying, and the changes kicked in just at the moment when the overall economy was tanking as well. For his part, Hicks doesn't actually recommend jettisoning the minimum wage, but he does suggest having a lower minimum wage for students and new hires. Still, other studies have found different results. When New Jersey raised its minimum wage in 1992, Princeton economists examined the impact on fast-food jobs there compared to Pennsylvania, where the minimum wage stayed the same, and found little difference.[25]

Another important argument, one sometimes raised by progressives as well as conservatives, is that raising the minimum wage doesn't really help poor families, which is the point for many advocates. Looking at figures from 2003, researchers from the Economic Policy Institute found that fewer than one in five minimum wage earners actually lives in a household below the poverty line, and only 9 percent are the head of the household,[26] like the single mother cited above. Minimum wage jobs are sometimes held by students or as second jobs, or the minimum wage worker is only one income in a household where other people also have jobs. Some progressives believe that there are better ways (such as the earned income tax credit) to target help directly to lower-income people.

In some states and the District of Columbia, the minimum wage is actually higher than the federal law requires—it's $8.50 in Oregon, for example.[27] As of July 2011, Oregon's unemployment rate was 9.5 percent, which is higher than that of most states.[28] But South Carolina, Georgia, and Mississippi had even higher unemployment rates, even though they just follow the federal law.[29] The bottom line is that raising the minimum wage is like a lot of other ideas we've covered here: it may affect the number of jobs available, but clearly there are other factors at work as well.

6. REDUCE BUSINESS COSTS BY REFORMING THE LEGAL SYSTEM TO REDUCE FRIVOLOUS LAWSUITS

According to the Chamber of Commerce's Institute for Legal Reform, "America's legal crisis is putting employees out of work, raising consumer prices, driving down shareholder value and bankrupting companies."[30] Six in ten small business owners say their businesses would be more profitable if they didn't have to fear "frivolous" lawsuits, and of those, almost two-thirds say they would use the extra revenue to hire people.[31] And litigation, many critics say, can be a job killer. According to a study conducted by Nobel Prize–winning economist Joseph Stiglitz, litigation over asbestos, a cancer-causing substance used in heavy industry and mining, has driven more than sixty companies into bankruptcy.[32] Although many workers were harmed by exposure to asbestos and clearly deserved help, the insurance industry maintains that over time the claims "increasingly and overwhelmingly [came] from healthy plaintiffs."[33] These are some reasons why a number of experts say it's time to reform the legal system so that only well-founded civil suits go to court. Iowa senator Charles Grassley, for one, has called for limits on class action lawsuits, where lawyers represent thousands of people in a case against a company or the government.[34] Critics charge that most of the money goes to the lawyers, not their clients, and in June 2011, the Supreme Court narrowed the definition for class action suits in a ruling for Walmart and against the plaintiffs, who represented some 1.5 million women employees.[35]

Other critics suggest going to a loser-pays system so people will think carefully before suing and so that defendants who aren't in the wrong aren't stuck with huge legal bills anyway.[36] The United Kingdom, Germany, France, Switzerland, and Spain all use this kind of system.[37]

Most Americans agree that unwarranted lawsuits are a problem, according to surveys,[38] but there are two main arguments against this approach. One is the possibility that by discouraging people from going to court, these changes will discourage people from suing even if they have been wronged.[39] But the other question, which is more important to what we're talking about here, is whether changes such as these really protect or create jobs. A *Duke Law Journal* analysis disputes the notion that the United States is "experiencing an escalating epidemic of litigation"—author Deborah Rhode points out that current U.S. litigation rates are not particularly higher than in the past or than the rates in other Western industrial countries.[40] A study conducted by the Economic Policy Institute concludes that legal reformers "inflate" the cost of lawsuits and "provide no credible evidence to support their assertions [that lawsuits are costing jobs]. In fact, what little effect changing the tort system will have on the economy might hurt job creation rather than help it."[41] Whatever its flaws, our current legal system does employ a lot of people. The United States has about 760,000 lawyers[42] and more than a quarter of a million paralegals.[43] At least some of those jobs might be lost if civil lawsuits were severely curtailed. Moreover, not all industries facing a lot of lawsuits are cutting jobs. Nearly everyone says the current medical malpractice system is broken, but health care is one of the few sectors in the economy showing rapid job growth. We might want to fix the legal system and reform malpractice law because it needs to be fixed, but should we count on this strategy to create jobs?

7. PROMOTE JOB SHARING AND UPGRADE PART-TIME WORK

One way to create more jobs would be to split the work in half and hire two employees half-time, rather than just one person full-time, to do the job. Many economists, including the influential *New York Times* columnist Paul Krugman, see this as a much better way to handle layoffs when a company just can't afford to keep all its employees.[44] Some states encourage job sharing by covering some of the wages that are lost when one salary is basically stretched out to cover two people. In 2010, between 300,000 and 350,000 workers were in job-sharing arrangements, which some experts believe saved about 100,000 jobs.[45] Job sharing could work several ways. Some proposals involve government funding: even though employees work less, their incomes do not drop precipitously. Others mainly focus on encouraging employers to offer this option to employees who want to work less. Human resource specialists say there are many people (such as parents with young children or older workers who want to cut back but not retire fully) who would actually prefer to work fewer hours.[46] Other analysts argue that American workers might benefit from longer vacations, better hours, and earlier retirements. This would spread the work out among more workers, both reducing unemployment and adding to our quality of life.

There are several objections, and we bet you're already on to some of them. Working half-time might be better than losing your job, but most people aren't exactly ending the month with a lot of Benjamins left over. Maybe this would be a useful temporary, stop-loss measure, but the people whose salaries are being divided up may not see it as a great solution, except in an emergency. Paul Krugman's support for the idea prompted one professor to suggest that the *New York Times* "reduce Mr. Krugman's presence on the page to, say, one column per year. The remaining hundred or so columns . . . could be written by unemployed economists."[47]

Indeed, a key criticism is that job sharing doesn't really create any of the *new* jobs that we need to make up for the jobs lost in the Great Recession. Former White House economist Lawrence Summers points to this drawback: "It may be desirable to have a given amount of work shared among more people," he has said. "But that's not as desirable as expanding the total amount of work."[48]

8. REDUCE AND REFORM CORPORATE TAXES

For *New York Times* columnist David Leonhardt, the United States now has "a corporate tax code that's the worst of all worlds. The official rate is higher than in almost any other country, which forces companies to devote enormous time and effort to finding loopholes. Yet the government raises less money in corporate taxes than it once did, because of all the loopholes that have been added in recent decades."[49] So there's a growing chorus of voices in both political parties calling for reducing the corporate tax rate from the current base level of 35 percent (which is higher than in many other countries) but eliminating many of the special tax breaks that allow some businesses to pay much less than others. For example, despite the high overall rate, Boeing only paid a total tax rate of 4.5 percent and Yahoo! 7 percent over five years, according to a *New York Times* analysis.[50] According to proponents, this would make U.S. business more competitive internationally and would entice more foreign corporations to open up U.S. subsidiaries that would hire people here. And for anyone who's upset by the idea of reducing taxes on multimillion-dollar enterprises, some studies suggest that most of the burden of the taxes is falling on workers anyway—that is, the businesses take most of the money for taxes out of worker salaries.[51]

The drawbacks? First, we always have to worry about the country's routine federal deficits and mushrooming debt. According to the Congressional Budget Office, corporations will pay about $400 billion a year in taxes over the next decade,[52] and if corporations pay less, someone else will need to pay more, or the red ink will keep on splashing. Corporate taxes are relatively low historically,[53] and the lower rates certainly haven't guaranteed strong job creation. What's more, a 2008 analysis by the CBO concluded that reducing corporate taxes does "not create an incentive for them to spend more on labor or produce more, because production depends on the ability to sell output."[54] In other words, the CBO suggests, corporations make decisions about hiring based on sales, not on what they have to pay in taxes.

9. REDUCE REGULATION BY HAVING LAWS "SUNSET" AND SHARPLY REDUCE THE ONES THAT APPLY TO SMALL BUSINESSES

There are those who argue that government regulations have gotten too complex and burdensome, especially for small businesses. President Obama issued an executive order ordering federal agencies "to review and remove outdated regulations and, where consistent with law, consider the costs and benefits of a regulation and choose the least burdensome path."[55] The Heritage Foundation has gone further, calling for regulations to "sunset," or expire, after a certain period of time. The regulations would have to be reinstated by Congress, so there would be a routine weeding process to eliminate those that didn't maintain broad political support.[56] Heritage also suggests strengthening the Office of Information and Regulatory Affairs (OIRA), a federal agency that determines whether new laws are cost-effective.[57] The impact of regulations on small businesses is

a special concern. One study by the Small Business Administration suggested that even though laws frequently provide exemptions for small businesses, companies with fewer
than twenty employees spend about $10,000 per worker
complying with federal regulations of different sorts.[58]

As an idea, hardly anyone is opposed to this, but other
experts say that businesses routinely overestimate what regulation really costs them[59] and that critics routinely underestimate the benefits to society in terms of clearer air, safer
workplaces, and so on.[60] What's more, some regulations
create jobs—companies have to hire people to make sure
they're obeying environmental laws, for example. Remember the number we cited earlier about the number of special
education teachers in the United States (473,000).[61] Those
jobs didn't exist before Congress passed laws requiring
schools to teach children with special needs.

Of course, like many ideas related to government, this
one seems to be easier said than done. OIRA, the agency
that's supposed to check on whether regulations are cost-
effective, is more than thirty years old, and it hasn't been
able to prune federal regulation much so far.

10. ENCOURAGE MORE PEOPLE TO START BUSINESSES

The United States has been losing jobs in many ways—in
fact, to use a more poetic turn of phrase, let us count the
ways. We lost some in the recession. Some have gone offshore. Some have been eliminated by technology. One of
the best ways to turn the tide, according to some experts,
is to get more new businesses up and running. It's new
businesses (as opposed to small businesses) that are most
likely to hire people. Unfortunately, the number of business start-ups has dropped by about 17 percent since the
Great Recession.[62] One study showed that even though

new businesses only account for about 3 percent of overall employment, they provide nearly a fifth of new jobs.[63] Many experts say the only way the United States can begin to catch up is to get more new businesses started. *Inc.* magazine—which is directed at entrepreneurs—compiled a list of sixteen different ideas for encouraging more start-ups, including cutting the red tape, providing microloans, upping government funding for science, allowing more immigrants who have the skills to start a business in this country, and passing an energy bill (right now, entrepreneurs in energy aren't sure whether the government is going to tax or regulate fossil fuels, so they don't know how to plan, and investors don't know where to put their money).

Like so many of the ideas here, nearly everyone likes the overall goal. The United States prides itself on encouraging people to pursue their dreams, and encouraging entrepreneurs to start new businesses certainly falls into that category. But the specific pieces aren't as popular. There is very little public support for expanding legal immigration,[64] and elected officials have steered away from reforming our current system because public opinion is so divided.[65] The same is true for energy—environmentalists want to encourage alternatives by taxing or regulating fossil fuels, but many existing businesses say this will just drive up their energy costs. And with federal and state governments under such financial stress, spending tax dollars for scientific research and providing loans for start-ups is not as easy as it once was.

11. MAKE HIGHER EDUCATION MORE AFFORDABLE AND MORE ATTUNED TO JOBS AND JOB CREATION

Many experts believe that the United States needs more college-educated workers if U.S. companies and our economy are to be competitive internationally. But getting more people to finish college is tough when the cost keeps going up. Based on a 2006 analysis from the National Center for Public Policy and Higher Education, median family incomes have risen by about 6 percent over the last decade, but tuition costs at public universities have jumped 44 percent.[66] In a recent Public Agenda survey asking Americans what would help them most in today's difficult economy, keeping college affordable was right at the top.[67] So the idea of encouraging higher education to hold the line on costs is attracting a lot of attention among policy makers. Some ideas include cutting administrative expenses, consolidating programs in state university systems, using more online courses,[68] and offering more one-year programs specifically targeted to fields where there are worker shortages (such as health care, accounting, or modern manufacturing).[69] Other experts say higher education itself should be more business-oriented. The Ewing Marion Kauffman Foundation, for example, has training programs to promote entrepreneurship among graduate students in the sciences and technology. The idea is to get them better prepared to start their own companies when they complete their degrees.[70]

Most Americans see getting a college degree (or having their kids get one) as part of the American dream, and college-educated people generally have lower levels of unemployment and better pay. But that doesn't necessarily mean that sending more people to college and/or retooling higher education to make it more jobs-oriented will do much to create more jobs anytime soon. As we point out in our chapters on technology and globalization, some say that the "college advantage" is actually disappearing as technology makes it easier for all kinds of jobs to be done anywhere in the world, even those requiring college degrees. Paul Krugman argues that "the notion that putting more kids through college can restore the middle-class society we used to have is wishful thinking. It's no longer true that having a college degree guarantees that you'll get a good job, and it's becoming less true with each passing decade."[71] Plus many well-educated societies stumble economically. Western European countries have highly educated populations, but many still struggle with very high unemployment rates, especially for the even better-educated younger people. According to one study by the United Nations, an oversupply of college-educated young people in industrialized countries is leading to "qualification inflation," where employers look for more credentials regardless of whether the job really requires them.[72] Argentina has one of the best-educated populations in Latin America,[73] but its professionals fled the country when the economy tanked after a debt crisis. So while most experts say that having a well-educated population is undoubtedly a good thing, there's much more disagreement about whether sending more people to college will, in and of itself, pull the American economy out of its doldrums.

12. SUPPORT THE UNION MOVEMENT

The goal of business is not to create jobs and pay good salaries. Its goal is to earn profits and give investors a return on their investment. This fact is at the core of the free enterprise system, which has allowed the United States to prosper since the nation was founded. But to offset the interests of owners and make jobs and better salaries a higher national priority, some experts say it's time to strengthen our unions.

As authors Jacob Hacker and Paul Pierson put it, unions are "a counterweight to the demands of corporations and Wall Street in the corridors of power."[74] They can fight layoffs and decisions to move jobs offshore. In some instances, they can create jobs for their members by requiring that employers operate in certain ways. Unions representing musicians typically negotiate contracts with Broadway producers requiring that they hire a certain number of live musicians rather than use recorded music.[75] Unions can also ensure that when companies are profitable, it's workers who benefit, not just owners and shareholders. According to former secretary of labor George P. Shultz, unions are essential in free market economies because they offer "a system of checks and balances."[76] What's more, union supporters say, having robust unions is entirely consistent with having a thriving economy. Germany has strong labor unions, but unlike the United States, it survived the global economic downturn with only minor increases in joblessness.[77] To strengthen unions, supporters say, we need to pass laws that protect union organizing and collective bargaining and oppose measures that weaken it.

But, as always, there's another side to the story. Critics charge that unreasonable demands for wages and better work rules by unions have put U.S. businesses at a real disadvantage in the global economy. Rather than helping job creation, unions stymie it, critics say. Their demands undercut companies' ability to invest in new products, expand their businesses, and hire more workers. Union may protect some jobs, but only for their own members, especially those with seniority.[78] Since unions (like many other groups) support political candidates, many critics see them as a special interest that promotes legislation benefiting union members but harming the economy overall.[79] The Manhattan Institute's Steven Malanga points out that "labor has been winning legislative victories for years at the state and local level where they've been able to pass bushels of labor-friendly laws," such as laws that make it difficult for nonunion companies to get contracts or easier for unions to organize. Surveys suggest only about one in five Americans has a "very favorable" opinion of unions,[80] and union membership has been declining for decades. To critics, this means that trying to strengthen unions is exactly where we don't want to go. Unions may have played an important role in protecting workers in the past, critics say, but in a modern global economy stronger unions will not be a plus.

13. GET BUSINESSES OUT OF THE HEALTH INSURANCE BUSINESS

About six in ten Americans under sixty-five get their health insurance through their job,[81] but as business writer Geoff Lewis points out, the arrangement is "more historical accident than public policy."[82] Companies began providing health insurance during World War II, when wages were frozen and labor was scarce, and after the war, more and more companies followed suit.[83] The system is costly for employers. In 2010, they paid nearly $10,000 per employee for family coverage, twice as much as they paid just a decade ago.* Many analysts say that having business pay for health care makes U.S. companies less competitive internationally and eats up money that could be going into research and development, business expansion, adding to the workforce, and improving salaries. Plus the system leaves people whose employers don't offer health insurance, who don't

---

* The average family plan cost $13,777, of which the employer covered $9,773 and the employee paid $3,997. Kaiser Family Foundation, Employer Health Benefits 2010 Annual Survey, Summary of Findings, 1, http://ehbs.kff.org/pdf/2010/8086.pdf.

work, or who only work part-time out in the cold.

So maybe it's time to find another way to provide health insurance and get employers out of the middle of it. Depending on your political leanings, that might be a system where everyone buys his or her own health insurance, with a tax write-off to help defray the costs. (The Heritage Foundation has outlined something along these lines at www.heritage .org/Initiatives/Health-Care.) Or it could be something like Medicare, except that instead of just covering seniors, the government insures everyone (doctors, hospitals, and so on would remain private, just as in Medicare). Physicians for a National Health Program at www.pnhp.org has suggested this approach. The idea here is to move away from a system that lays health care costs on businesses, which dampens business expansion and weighs down job creation. From the left and the right, there are those who say that business's time and money could be better spent conducting business—not selecting and managing health insurance policies for their employees.

We don't have to tell you how controversial this all is, and for a lot of Americans, the idea of having another political debate on health care is enough to cause a headache itself. What's more, polls suggest that rather than wanting to jettison the employer-based system, most Americans want it expanded. Seven in ten Americans say the government should require businesses to offer health insurance to full-time employees.[84] In the end, a lot of analysts just don't think it's politically possible to move to a different system at the moment, either one based on individuals buying their own plans or one in which the government is the single payer.[85] So for now, one of the most persuasive arguments against this strategy is that it would be a huge diversion of time and effort that could be better spent on other ideas to create more and better jobs.

14. CREATE A LEVEL PLAYING FIELD WITH OUR TRADING PARTNERS

Globalization basically means that "it's a small world after all," and that's not going to change. But we can do something to make sure the global market works for everybody. Critics of globalization argue that many countries in the developing world are more than eager to pay low wages, ignore pollution, and subvert workers' rights in order to attract business. And major Western corporations are more than willing to go along with this. The answer, some economists and activists argue, isn't to build barriers against foreign goods or offshoring jobs, but to use our trade policy to make sure other countries treat their workers better. We can demand, as part of our trade agreements, that any country trading with us meet international standards on union organizing, child labor, minimum wages, and environmental protection. The other countries don't have to agree, but if they don't, they don't get the trade deal. In fact, there was a bipartisan agreement to get behind that idea in 2007, when the Bush administration and congressional Democrats agreed that future free trade agreements would include labor standards.[86]

Supporters argue that if developing countries followed these rules, not only would their workers benefit, but a lot of their unfair cost advantage over American workers would evaporate. Ideally, it's better for everybody. "Standards of living converge on an upward trajectory," said John D. Podesta and Sabina Dewan of the Center for American Progress, which has a "Just Jobs" initiative in partnership with other think tanks internationally. "Rules and regulations ensure this happens across the globe. And the resulting economic and political stability rebounds to benefit developed and developing countries alike."[87] Another way of attacking this issue is the fair trade movement—you may have seen the label on products in the supermarket, like coffee. Fair trade is a process where participating businesses agree to follow a set of standards in dealing with producers in developing countries, including transparent business practices, meeting international labor standards, and paying fair prices.*

Critics say it's the job of other governments to protect their workers, not the job of the United States, and also point out that global trade is already improving living standards in other countries—maybe not as fast as we'd like, but it is happening. The wages paid by new factories in developing countries may be far below what Americans make, but it's still a raise over what people were making before. They get a rising standard of living; we get products at a discount. The libertarian Cato Institute says fair trade organizations are well-intentioned and have helped many participating producers, but also attempt to "strong-arm the market" by trying to set a price floor on coffee.[88]

There are also those who doubt how effective this would be. Trade agreements can make a difference in labor markets, but other factors, such as the labor supply, education levels, and available capital to invest, matter more. Plus, the wage gap with foreign workers can be so great (for example, Mexican workers make only 11 percent as much as Americans on average) that even stronger unions and higher standards may not close the gap.[89]

This strategy assumes that you can get most countries, or at least most of the major countries, behind it. The United States is the world's largest economy, so we have a lot of clout on this, but if enough other industrial nations refuse to go along, this probably isn't going to work. And we

---

* You can find out more at the World Fair Trade Organization, www .wfto.com, and see their 10 Standards for Fair Trade.

must not assume this is going to be a fast fix. It took nearly a decade for the United States to negotiate the 1994 North American Free Trade Agreement with Canada and Mexico, and for Congress to pass it.[90] It is still contentious more than a decade later.[91]

THE LONG AND THE SHORT OF IT

Ever since the U.S. economy took a nosedive in late 2008, elected officials and political candidates have been floating all sorts of ideas for preserving and creating jobs. One major consideration that a lot of us non-economist, non-policy-wonk types sometimes overlook is whether a proposal is aimed at jolting the employment picture for a short period of time to help the country through a recession or whether it's a policy change aimed at strengthening the job picture over the long haul, say the next decade or two.

Helping the country through economic tough times is crucial, especially in addressing the cyclical job loss that's mainly caused by a poor economy. So are ideas that aim at improving the U.S. jobs situation over time—we really need both. The key thing here is having reasonable expectations. We can't expect short-term cures to fix long-term problems such as job losses due to globalization or technology. And we can't expect long-term solutions to kick in quickly.

Here's a chart with some much-discussed examples to consider.

IDEA	SHORT TERM	LONG TERM
Have the government jump-start major building projects	Yes and no. If the projects are shovel-ready (everybody is on board with the plans and the money), this can create jobs quickly. But if it's not, it can take years to get all the approvals and raise the money, so hiring can be delayed.	Obviously, building and construction projects don't provide jobs forever, but major national infrastructure improvements such as upgrading the electric grid or developing wind or solar power networks can take decades. Plus, they lay the groundwork for economic development in the future.
Improve K-12 education	Near-term jobs for teachers and child development professionals.	This is mainly a long-term strategy. Down the road, a better-educated workforce is more likely to produce the innovation and productivity increases that lead to more jobs.
Extend unemployment benefits	Surprised to see this here? In the short term, this does seem to help preserve and create jobs. It gives unemployed people an income. Then they're able to spend at local businesses, and those local businesses are less likely to lay off workers and more likely to hire. The Congressional Budget Office gives this strategy a pretty good rating for helping on jobs.	This is mainly a short-term strategy. The idea of supporting people who aren't working for a very long period of time has some important downsides. It can lead people to be unrealistic and refuse jobs they probably should take, and over time it might become very divisive. One reason the welfare system was so deeply unpopular in the 1980s was that it was possible to stay on welfare for years, and the people working and paying taxes resented it.
Cut income tax rates	May give people more money in their pockets, which bolsters consumer spending and helps merchants, possibly preserving existing jobs. According to the CBO, this is not as helpful in creating new jobs quickly as some other approaches.	May encourage wealthier Americans to invest in new ventures or buy stock, which could mean more jobs in the years ahead.

Cut payroll taxes	Almost certainly a short-term idea, and, according to studies by the Congressional Budget Office, one of the most effective ways to persuade employers to hire.	Not a long-term solution. Since both Social Security and Medicare face serious long-term funding problems, not collecting these taxes for any substantial period makes the shortfalls even worse.
Pursue more trade agreements	This may not create jobs tomorrow, but opening up foreign markets for U.S. products can lead U.S. companies to beef up their workforces to meet the new demands.	Most economists believe that trade is good for jobs over the long haul, but it can be a mixed bag. Some companies may lose business, reduce their workforces, or even close due to foreign competition, while others expand and need more workers. The main issue here is that not trading with other countries could turn out to be even worse for jobs.
Invest more in scientific research and develop cutting-edge technology	It's mainly a long-term solution, although it certainly does provide jobs for researchers and college professors in the near term.	Over time, this can provide the innovations and scientific breakthroughs that open entire new industries, creating thousands of jobs. But it doesn't happen overnight, and it's not always clear what's going to produce that eureka moment.
Refurbish houses and buildings to make them more energy efficient	Mainly short-term. There's some organizing involved, but the jobs don't require huge amounts of training, and the projects don't require huge amounts of money or a long approval process, so investment could lead to jobs reasonably quickly.	Once you've done the caulking and added the insulation, you're set for a while. On the other hand, the United States has about 77 million freestanding houses, and only about 1 million of them are up to Energy Star standards, so this could create jobs for a while.

||||||||||||||||||||||||||||||||||||||||||||||||||||||||||||||||||||||||||||||

CHAPTER 17

NOW WHAT? SEVEN GROUND RULES FOR MOVING FORWARD

One thing is sure. We have to do something.
We have to do the best we know how at the
moment.

—*Franklin D. Roosevelt[1]*

In this book, we've pursued what we hope is a useful mission—explaining the decisions the United States faces on jobs, trying to avoid economic jargon, and focusing on the practical questions rather than big theoretical and ideological ones. We've done our best to gather the most reliable data and suggest where the experts are mainly in agreement and where they're not.

By now, we suspect we've aggravated sufficient numbers of economics professors and policy wonks who think we've oversimplified the issues. We'll plead guilty as charged on that, but we are trying to help people catch up so that they can actually understand all those finer points and subtleties. On the other hand, maybe we've frustrated some of you who are looking for clear, guaranteed answers on how to get the country's job creation machine running full steam ahead. Unfortunately, based on the research, we don't think anyone has surefire answers. Despite the campaign slogans you're likely to hear in election season, there just aren't any simple solutions, and there certainly aren't any that are cost-free.

Now that we're about to wrap up, we're still not going to recommend any specific policies or solutions. We believe sorting through the options and deciding what's best for the country is the nation's collective job now. We have to do a better job scrutinizing candidates for office and being more thoughtful and skeptical about their ideas on jobs. Then we'll need to band together as citizens and neighbors to support them and/or keep the pressure on once they've taken their oaths as leaders at the federal, state, and local level.

But just because we're not going to promote a specific set of solutions doesn't mean we don't have some ground rules for you to keep in mind while you're thinking about what's worth trying and what's likely to work. Here are some broad ideas that we think are key for you—and all of us together—to find a better path for addressing our country's disappointing jobs situation. Here goes:

The president and Congress can't just "fix it."

In the next election and for many elections to come, jobs will be—and should be—a top issue. But we need to be honest with ourselves about what politicians can actually do to create jobs and help us hold on to the jobs the country has. Laws, policies, and regulations provide a context for job creation. We've spent pages describing different ways to think about that, and what government does is important. But in our economy, most jobs are created in the private sector, and in our form of government, politicians can't actually make companies hire people. We hope elected officials will make some good judgments about policies that will help businesses thrive and hire people (here in this country) and make some equally good judgments about when to get out of the way and let the private sector do its thing. But don't give any politician your vote thinking that he or she will be able to deliver more and better jobs just by taking over the reins of office. The most they can do is nudge things in the right direction.

Jobs is the word—it's got groove, it's got feeling.
We'll borrow Frankie Valli's lyrics from *Grease* to make a key point:[2] the United States is not going to be very successful at solving the jobs problem unless we actually talk about creating and holding on to *jobs*. And that means more than merely tossing the word into political speeches as a poor cousin of *economic growth* or *prosperity*. We can all agree that the country needs economic growth—even Mr. Spock was always telling people to "live long and prosper"—and we won't get jobs without it. But the truth is that economic growth is not the same thing as having more and better jobs. The world has changed, and our economy has changed. Now, companies can build their businesses by relying more on technology to do jobs that used to be done by humans. And they can grow and become more profitable by doing more business abroad. Those things are fine—and we've mounted spirited defenses of both technology and globalization in this book. But those facts also present us with a challenge. How are we going to spur job creation right here in the USA—good jobs for our people? Aiming for economic growth is terrific, but we also need to talk about jobs consciously and specifically.

Think like an employer.
We mentioned this piece of advice earlier, and it bears repeating. In fact, we think it's so important to solving

the country's jobs problems that maybe we should put it on mouse pads and coffee mugs. People get hired when a human being running a company or other organization needs additional human beings to help get the work done—we'll even count individual human beings who hire people to help with everything from filing their taxes to organizing their closets. So what we need to do is create an environment where that's more likely to happen. Whether we're talking taxes or government regulation or education reform, we must keep that end result clearly in mind: will this make it more or less likely that a real person somewhere will hire someone and pay that individual a decent wage, or is it going to make hiring people so much more complicated and expensive that, from the employer's point of view, it's just not worth the money?

We're not saying American business should get anything and everything it wants. Businesses exist to create profits, not jobs; to businesspeople, jobs are a means to that end. But most Americans also want to make our country safer and fairer. And some of the changes we make in those directions create jobs as well. But we do need to keep our eyes on the "think like an employer" ball. Despite all the "job creation" and "job-killing" verbiage in Washington and elsewhere, there's way too little down-to-earth discussion of what actually leads employers to hire people.

Don't expect employers to do things that will sink their businesses.
Some ideas that seem like good ways to save jobs or create new ones are like shots of tequila—the immediate effect seems wonderful, but go too far and you'll be as sick as a dog. When employers lay workers off during recessions, when they embrace technology that eliminates jobs, or when they open factories in other countries, often our first

impulse is to call for the sheriff and try to stop them. We should expect employers to be good citizens and to be loyal to their workers and their communities, but we can't expect them to operate in ways that pull their company underwater. There needs to be a happy medium here.

There's some good evidence that being too quick to lay people off can backfire on employers—it makes investors think their business is rocky, and they lose hard-to-replace talent and have to spend money to get that talent back later. But over the long haul, employers have to make decisions that make sense for their number one priority: ensuring that their business is financially solid and produces a profit. We need to accept that. That doesn't mean we can't ask businesses to take steps that benefit the community and the country as a whole. And it's certainly fair to ask them to play by rules that make the country better—keeping the air clean, for example.* But if we're excessive in what we ask of employers, the end result could be devastating. The United States needs healthy, prosperous businesses, or we won't have enough jobs.

Let's start stuff.

Economists are duking it out over how to get the economy moving, and the researchers and statisticians can leave you punch-drunk with all their arguing over whether our

---

* In fact, people who head businesses often hate indecision more than they dislike dealing with new laws and regulations. When *Inc.* magazine compiled its wish list for spurring entrepreneurship, it said: "Markets—and investors and entrepreneurs—abhor uncertainty. So let's get serious about the emerging energy economy by creating an actual energy policy. Only then will companies be able to make informed investment decisions." "Revitalizing the American Dream," *Inc.*, July 6, 2010, www.inc.com/magazine/20100701/revital izing-the-american-dream.html.

standard of living is improving and whether having more people go to college actually helps create jobs (among other things). You would think that someone actually knows the answers to questions such as these, but still the battle rages. In fact, in doing this book, we've been surprised at just how many aspects of the jobs issue are in dispute. At the same time, there does seem to be consistent information showing that most jobs creation comes from new businesses, rather than older, more established ones—which makes sense. There's also good information suggesting that inventing things and coming up with better ways of doing things that other people will pay for is a key to keeping the economy vibrant and businesses hiring. We're not going to say that new is always better, but it does seem to give a kick start to hiring.

And let's keep the door open.
The world is getting smaller, and borders mean less than they used to, which is frightening to a lot of people on several fronts. There are lots of people who'll tell you that jobs going overseas or people immigrating here to work are inherently bad things and we have to make them stop. But the data don't bear that out. For every negative around globalization and immigration, there's a positive, and arguably the positives contribute to the U.S. economy, as opposed to undercutting it.

A lot of people want to come here because it's a place where they can start new businesses, educate their children, and work their way up in the world. We need to welcome people like that, and we especially need to welcome people from other countries who excel in science, math, engineering, inventing, entrepreneurship, sales, and such. Maybe there'll be some point in the future when we have too many people like this, but it's hard to see when that might be.

There is some legitimate worry that increased immigration of low-skilled workers may hold down pay for low-skilled American-born workers, but even here the evidence is very mixed, and there may be better solutions than closing the door on immigration. It's certainly legitimate for anyone living in the United States, whether born here or not, to be concerned about immigration policies that aren't enforced and people wandering in and out of the country in defiance of our laws. That's not good. But having people come here legally to join us in building up our economy is something we ought to continue.

It's also tempting to try to rein in globalization, or at least take tougher steps to try to stem the tide of U.S. jobs going abroad. There's no question that a lot of jobs, even high-skilled jobs, are no longer tied to a specific place. We're competing with educated people all around the globe. But, realistically, we can't make this process stop. Even if we raised trade barriers and imposed heavy penalties on companies that shifted jobs overseas, it's hard to see how the world's response would be anything other than raising barriers against us in response. The world would continue globalizing without us, and we might end up a backwater nation.

Some kinds of work will be done cheaper or even better elsewhere, no matter what we do. The trick is going to be adapting to change: finding the work that we can do better than anyone else, and ensuring that people who are unmoored by this change get all the help they need to rebuild their careers.

It's time to drop the tunnel vision.

One of the points we've tried to make repeatedly is that jobs come from lots of different kinds of businesses (including government), and that countless factors go into creating a good climate for keeping good jobs and generating new ones. And despite all the huffing and puffing on Capitol Hill and CNBC, no one really has the magic formula. So one extremely beneficial change would be for politicians and pundits to stop acting as if any one thing—tax policy, or reducing red tape, or promoting unionism—is the killer app for solving our jobs dilemma.

Sometimes it seems as though the whole jobs problem is like a supernova that collapses in onto itself to become a white dwarf of a debate over taxes. Taxes are important. We don't want taxes crushing businesses and making it tough for them to hire, and we don't want the government so impoverished that failing schools and decaying infrastructure undercut businesses either. Tax policy certainly matters, but in the end, businesses won't hire people unless they need them to do their work, and if they really need people to help them get the work done, taxes alone won't stop them. Let's have a sensible debate on taxes, but we need to stop pretending that it's the be-all and end-all for jobs and job creation.

So let the games begin. It's time to make some decisions about the best ways to keep the jobs we have, get more of them, and get more of the ones that pay people a decent wage. And if something's not working, let's be open-minded and spirited enough to try something else.

WORKING HARD OR HARDLY WORKING

Sometimes people from other countries sneer at Americans because most of us do seem to like to work; many of us seem to like starting businesses and being our own bosses, and just about all of us like earning money. Maybe we don't spend enough time stopping to smell the roses, but work is a key part of our national ideal—if you're willing to work hard, you can build a good and decent life for yourself. The question we face now is whether the U.S. economy is providing enough opportunity to make that possible for all of us.

Even before the Great Recession, there seemed to be a greater insecurity about work in America. You could see it in our pop culture: when Hollywood delves into the workplace, the main thing it finds is insecurity.

Think, for example, of *Glengarry Glen Ross*, the David Mamet movie about the cutthroat world of a real estate office, which opens with star salesman Alec Baldwin bellowing at the others, "Put that coffee down! Coffee is for closers only!" The rest of the movie is a desperate struggle among the salesmen who know that the first prize in the monthly sales contest is a Cadillac, second is a set of steak knives, and third prize is being fired.

Or take *Office Space,* the cult comedy about life in one of those rectangular cubicle farms you can drive past on nearly every highway in America. The workers there live with a constant stream of petty humiliations over filing their mysterious "TPS reports" and under threat of being axed by consultants, all of whom are named Bob. They take their lunches at a chain restaurant where the waitstaff are harassed by the manager for not wearing enough faux-funky buttons and stickers, referred to as "pieces of flair." Yet despite this, all of them (with one notable exception) are desperate to justify their jobs, leading one employee to end up screaming at a couple of the Bobs, "I have people skills! Can't you understand that! What the hell is wrong with you people?"

And these are educated, white-collar workers, the people who should have the greatest advantages. We've talked often about "churn" in the labor market, and why it's a good thing on the whole, but in these stories, people are far more concerned with holding on to what they have than with getting a better job. Perhaps the worst-case scenario here is Michael Douglas in *Falling Down*, a tightly wound aerospace engineer, laid off and estranged from his family, who goes on a shooting rampage across Los Angeles, the first prompted by poor customer service.

Yet part of the frustration these movies tap into is that most people want work, and they want work that's satisfying to them. Even in the depths of the Great Recession, a Gallup poll found that nearly all employed Americans say they're at least "somewhat" satisfied with their jobs. In 2009 and 2010, despite unemployment above 9 percent, Gallup found eight in ten satisfied with their job security (about half were "completely satisfied"). They're less happy about how much they earn or what their benefits are, but on the whole, most working people are reasonably happy with their actual work.[3] That doesn't preclude problems—four in ten Americans say they're "struggling" economically, in a 2011 Public Agenda survey, and they're deeply concerned about their long-term prospects for retirement and paying for their children's education.[4]

But generally speaking, the job they have doesn't bother Americans as much as the possibility that they won't be able to find another one. In June 2011, fully eight in ten Americans told Gallup they thought it was a "bad time to find a quality job."[5] That's not surprising. What's more surprising is that at least half the American public has said this consistently for the past decade—between 2001 and 2011, the most optimistic point was in 2007, when the public was evenly split: 48 percent said it was a good time to look for a job, and 47 percent said it wasn't.[6]

If you've never seen the film *Kramer vs. Kramer*, it's about a father (Dustin Hoffman) who's involved in a divorce battle with his wife (Meryl Streep) and trying to keep custody of his young son. At one point in the movie the father loses his job, and he desperately needs a new one—if he's not working, he's lost any chance he ever had of gaining custody. He's an advertising art director, and he goes to apply for a job at a big New York City agency, arriving in the middle of the annual Christmas party. It's an uncomfortable situation, but he won't be put off. He uses every ounce of persuasion, resolve, and persistence he has to get himself hired, and it works.

The scene is wonderful moviemaking, but it's also a reminder that part of the solution to the jobs problem lies with us, the workers. If we lose a job, we need to look for a new one tirelessly, and if we have one, we need to perform it well and conscientiously. But it's also a two-way street: for example, in *Kramer vs. Kramer*, there was a job opening the protagonist could fight for, and he did have the skills and experience to give his new employer good service in return for hiring him. There have to be jobs to fight for.

The real challenge we face is making sure we've done everything we can, as a nation, to ensure that people have those chances. Right now we're in a position where the tide may be running against people trying to make it in our economy, partly because of the worst recession in eighty years and partly because of long-term trends that are going to change the face of work.

We're not worried about Americans losing their taste for work or their creativity in coming up with new ideas, new products, and new ways of creating jobs. But we are worried about creating a society where the greatest number of people have the greatest possible chance to get ahead. Jobs aren't just what we do, they're who we are, and they're the foundation for much of the rest of our society.

If the majority of Americans are willing to put every fiber of their being into finding work, then we, as a nation, need to put every fiber of our being into making sure there's work for them to find.

ACKNOWLEDGMENTS

One of the great assets we've had in writing this book has been the support and advice of a wonderful group of colleagues. Ruth Wooden and Deborah Wadsworth, both of Public Agenda's board of directors, helped inspire this series of books and have been steadfast supporters on all our projects. All of our Public Agenda board members and colleagues have been a great help to us, particularly David White and Allison Rizzolo, who've helped us work through crucial technical challenges. The insight, inspiration, and faith in the American public of Dan Yankelovich, Public Agenda's cofounder, underlies all our work.

We would also like to thank Jenny Choi, our indispensable fact checker, and Nancy Hagar, who created our charts and graphs. Jennifer Tennant and Andrew Yarrow provided helpful feedback on early drafts of our manuscript. We have benefited from the intelligence and professionalism of Jud Laghi, our literary agent, and from the guidance and encouragement of Larry Kirshbaum. In addition, our editors at HarperCollins have been extraordinary. Matthew Inman was instrumental in shaping the early stages of *Where Did the Jobs Go?* while Stephanie Meyers and Peter

Hubbard brought their exceptional skill and judgment to the final version. We'd also like to thank our production editor, Lelia Mander, for all her help throughout. Finally, this book wouldn't have been possible without the support of our families and friends. We would especially like to thank Josu Gallastegui and Susan Wolfe Bittle for their unfailing love, patience, and encouragement.

NOTES

PREFACE

1. If you'd like a trip down memory lane on how the Great Recession unfolded, check out this timeline from the University of Pennsylvania's Lauder Institute: http://lauder.wharton.upenn.edu/pdf/Chronology%20Economic%20%20Financial%20Crisis.pdf.
2. U.S. Department of Labor, Bureau of Labor Statistics, TED: The Editor's Desk, "Payroll Employment in February, 2010," March 9, 2010, www.bls.gov/opub/ted/2010/ted_20100309.htm.
3. Congressional Budget Office, "Budget and Economic Outlook: Fiscal Years 2011 to 2021," January 2011, 2, www.cbo.gov/ftpdocs/120xx/doc12039/SummaryforWeb.pdf.
4. See, for example, Michael Spence, "Why the Old Jobs Aren't Coming Back," *Wall Street Journal,* June 24, 2011, http://online.wsj.com/article/SB10001424052702303714704576385863720618134.html.
5. Bureau of Labor Statistics, "The Employment Situation—July 2010," www.bls.gov/news.release/empsit.nr0.htm.
6. Bureau of Labor Statistics, "Employed Persons by Class of Worker and Part-Time Status," Table A-8, June 3, 2011, www.bls.gov/news.release/empsit.t08.htm.
7. Bureau of Labor Statistics, "Employment Characteristics of Families Summary," March 24, 2011, www.bls.gov/news.release/famee.nr0.htm.
8. Pew Research Center, "The Great Recession at 30 Months," June 30, 2010, http://pewresearch.org/pubs/1643/recession-reactions-at-30-months-extensive-job-loss-new-frugality-lower-expectations.
9. Ibid.
10. Bureau of Labor Statistics, "Frequently Asked Questions: How Large Is the Labor Force?" www.bls.gov/cps/faq.htm#Ques3.
11. National Bureau of Economic Research, "U.S. Business Cycle Expansions and Contractions," www.nber.org/cycles/cyclesmain.html.

CHAPTER 1: IT'S NOT OVER UNTIL IT'S OVER

1. Fiona Ng, "Extra: Greedy 'Friends' or Justified 'Friends'?" Hollywood. com, April 26, 2000, www.hollywood.com/news/EXTRA_Greedy_ Friends_or_Justified_Friends/31223.
2. www.tv.com/friends/the-one-with-five-steaks-and-an-eggplant/epi sode/373/recap.html.
3. www.tv.com/friends/the-one-with-five-steaks-and-an-eggplant/epi sode/373/trivia.html?tag=episode_header;trivia.
4. U.S. Department of Labor, Bureau of Labor Statistics, "Since the Start of the Recession in December 2007, Payroll Employment Has Fallen by 8.4 million," January 2010, www.bls.gov/news.release/emp sit.0_2052010.pdf.
5. U.S. Department of Labor, Bureau of Labor Statistics, "The Employment Situation—December 2009," www.bls.gov/news.release/archives /empsit_01082010.htm.
6. Economic Projections of Federal Reserve Board Members and Federal Reserve Bank Presidents, April 2011, www.federalreserve.gov/ monetarypolicy/files/fomcprojtabl20110427.pdf.
7. Congressional Budget Office, CBO's Economic Projections for Calendar Years 2010 to 2021, Table 2-A, January, 2011, http://www.cbo.gov/ ftpdocs/120xx/doc12039/EconomicTables%5B1%5D.pdf.
8. Calculated from Household Survey Data, U.S. Department of Labor, Bureau of Labor Statistics, "The Employment Situation—December 2009," 2, www.bls.gov/news.release/archives/empsit_01082010.htm.
9. Labor Force Statistics from the Current Population Survey, June 3, 2011, www.bls.gov/web/empsit/cpseea04.htm.
10. Ibid.
11. www.bls.gov/cps/cpsaat7.pdf.
12. U.S. Department of Labor, Bureau of Labor Statistics, "Spotlight on Statistics: Back to College," September 2010, www.bls.gov/spotlight/2010/college/home.htm.
13. Neil Irwin, "Aughts Were a Lost Decade for U.S. Economy, Workers," *Washington Post*, January 2, 2010.
14. Ibid.
15. Ibid.
16. Mark Thoma, "What Level of Job Creation Will Sustain Full Employment?" *Economist's View*, November 3, 2006, http://economistsviewty pepad.com/economistsview/2006/11/what_level_of_j.html.
17. www.popculturemadness.com/Music/Pop-Modern/1999.html; www.nbc .com/ER/remembered/cast.shtml.
18. Conor Dougherty and Sara Murray, "Lost Decade for Family Income," *Wall Street Journal*, September 17, 2010, http://online.wsj.com/article/ SB10001424052748703440604575495670714069694.html.
19. Neil Irwin, "Aughts Were a Lost Decade for U.S. Economy, Workers," *Washington Post*, January 2, 2010.
20. U.S. Census Bureau, Current Population Survey, Average Poverty Thresholds, 2009, www.census.gov/hhes/www/poverty/data/historical/index.html.

21. Conor Dougherty and Sara Murray, "Lost Decade for Family Income," *Wall Street Journal*, September 17, 2010, http://online.wsj.com/article/SB10001424052748703440604575495670714069694.html.

22. McKinsey Global Institute, "Changing the Fortunes of America's Workforce: A Human Capital Challenge," June 2009, www.mckinsey.com/mgi/reports/freepass_pdfs/changing_fortunes/Changing_for tunes_of_Americas_workforce.pdf.

23. Bureau of Labor Statistics, "The 30 Occupations with the Largest Number of Total Job Openings Due to Growth and Replacements, 2008–18," www.bls.gov/news.release/ecopro.t10.htm.

24. McKinsey Global Institute, "Changing the Fortunes of America's Workforce: A Human Capital Challenge," June 2009, Executive Summary, www.mckinsey.com/mgi/reports/freepass_pdfs/changing_for tunes/Changing_fortunes_Executive_summary.pdf.

25. "U.S. Mortgage Delinquencies Set Record," Reuters, September 21, 2009, www.reuters.com/article/idUSTRE58K29E20090921.

26. *The Atlantic*, "How a New Jobless Era Will Transform America," March 2010, www.theatlantic.com/magazine/archive/2010/03/how-a-new-jobless-era-will-transform-america/7919/2.

27. Speech by Federal Reserve governor Frederic S. Mishkin, "Monetary Policy and the Dual Mandate," April 10, 2007, www.federalreserve.gov/newsevents/speech/Mishkin20070410a.htm.

28. Congressional Budget Office, "Budget and Economic Outlook Fiscal Years 2011–2021," January 2011, www.cbo.gov/doc.cfm?index=12039.

29. Murat Tasci and Saeed Zaman,"Unemployment After the Recession: A New Natural Rate?" Federal Reserve Bank of Cleveland, September 8, 2010, www.clevelandfed.org/research/commentary/2010/2010-11.cfm.

30. Bureau of Labor Statistics, "How the Government Measures Unemployment," www.bls.gov/cps/cps_htgm.htm#why.

31. Bureau of Labor Statistics, "Employment Situation News Release," April 1, 2011, www.bls.gov/news.release/archives/empsit_04012011.htm.

CHAPTER 2: HAS AMERICA LOST ITS MOJO?

1. http://lyricsplayground.com/alpha/songs/s/spinningwheel.shtml.

2. Joe Holley, "Actor Jack Palance; in a Varied Career, Roles as Embodiment of Evil Stand Out," *Washington Post*, November 11, 2006, www.washingtonpost.com/wp-dyn/content/article/2006/11/10/AR2006111001813.html.

3. World Economic Forum, *The Global Competitiveness Report 2010–2011*, www3.weforum.org/docs/WEF_GlobalCompetitivenessReport_2010-11.pdf.

4. Central Intelligence Agency, World Factbook, "United States—Economy," www.cia.gov/library/publications/the-world-factbook/geos/ us.html.

5. Central Intelligence Agency, *World Factbook*, "China—Economy," https://www.cia.gov/library/publications/the-world-factbook/geos/ch.html.

6. See, for example, Chris Oliver, "China's GDP to Overtake U.S. by Early 2020s, Says Analyst," MarketWatch, April 23, 2009, www.market-watch.com/story/chinas-gdp-overtake-us-early.

7. Central Intelligence Agency, *World Factbook*, "Country Comparison: Exports," https://www.cia.gov/library/publications/the-world-factbook/rankorder/2078rank.html.

8. Central Intelligence Agency, *World Factbook*, "Country Comparison: GDP," https://www.cia.gov/library/publications/the-world-factbook/rankorder/2001rank.html.

9. OECD, "The World Economy: A Millennial Perspective," Table B-18: "World GDP, 20 Countries and Regional Totals, 0–1998 A.D.," www.theworldeconomy.org/statistics.htm.

10. "Growth in GDP per Capita,"*Factbook 2009: Economic, Environmental and Social Statistics*, http://titania.sourceoecd.org/vl=2904820/cl=25/nw=1/rpsv/factbook2009/02/03/02/index.htm.

11. Ibid.

12. Ibid.

13. Public Agenda, *One Degree of Separation: How Young Americans Who Don't Finish College See Their Chances for Success*, 2011, www.publicagenda.org/files/pdf/one-degree-of-separation.pdf.

14. National Center for Education Statistics, *Digest of Education Statistics*, Table 408: "Average Mathematics Literacy, Reading Literacy, and Science Literacy Scores of 15-Year-Olds, by Sex and Country: 2009," http://nces.ed.gov/programs/digest/d10/tables/dt10_408.asp.

15. Central Intelligence Agency, *World Factbook*, "Country Comparison: GDP," https://www.cia.gov/library/publications/the-world-factbook/rankorder/2001rank.html.

16. www.weatheronline.co.uk/reports/climate/Norway.htm.

17. Central Intelligence Agency, *World Factbook*, "Luxembourg," www.cia.gov/library/publications/the-world-factbook/geos/lu.html; "Qatar," www.cia.gov/library/publications/the-world-factbook/geos/qa.html.

18. Nancy Folbre, "Sick at Work," *Economix* blog, *New York Times*, February 10, 2010, http://economix.blogs.nytimes.com/2010/02/10/sick-at-work/#more-51557.

19. U.S. Department of Labor, Bureau of Labor Statistics, "Employee Benefits in the United States," July 28, 2009, www.bls.gov/news.release/ebs2.nr0.htm.

20. Congressional Budget Office, "Current Budget Projections: An Update," August 2010, Table 1-2, www.cbo.gov/ftpdocs/117xx/doc11705/BudgetProjections.pdf.

21. U.S. Government Accountability Office, "The Federal Government's Long-Term Fiscal Outlook: January 2011 Update," March 17, 2011, www.gao.gov/new.items/d11451sp.pdf.

22. Cited in Steven Malangam, "The Muni-Bond Debt Bomb," *Wall Street Journal*, July 31, 2010, http://online.wsj.com/article/NA_WSJ_PUB:SB10001424052748703999304575399591906297262.html.

23. Money-zine.com, "Credit Card Debt Statistics," www.money-zine

.com/Financial-Planning/Debt-Consolidation/Credit-Card-Debt-Statistics/.

24. Congressional Budget Office, "Economic and Budget Issue Brief: Factors Underlying the Decline in Manufacturing Employment Since 2000," December 23, 2008, www.cbo.gov/ftpdocs/97xx/doc9749/12-23-Manufacturing.pdf.

25. Ibid.

26. Ibid.

27. World Economic Forum, *The Global Competitiveness Report 2010–2011*, United States Profile, pp. 340–41, www3.weforum.org/docs/WEF_GlobalCompetitivenessReport_2010-11.pdf.

28. Ibid.

29. Scott Bittle and Jean Johnson, *Who Turned Out the Lights? Your Guided Tour to the Energy Crisis* (New York: HarperCollins, 2009), 247, www.publicagenda.org/whoturnedoutthelights.

30. International Energy Agency, "World Energy Outlook, 2010: Presentation to the Press," November 9, 2010, www.worldenergyoutlook.org/docs/weo2010/weo2010_london_nov9.pdf.

31. www.songfacts.com/detail.php?id=3035.

CHAPTER 3: JUST THE FACTS, MA'AM

1. www.manifestyourpotential.com/work/game_of_work/choose_work_goal/quotes_work_jobs_careers_goals_inspirational.htm.

2. www.youtube.com/watch?v=A9CxT48jIgI, accessed April 22, 2011.

3. Bureau of Labor Statistics, "Employment Situation Summary," June 3, 2011, www.bls.gov/news.release/empsit.nr0.htm.

4. Bureau of Labor Statistics, "Civilian Labor Force Participation Rates by Age, Sex, Race, and Ethnicity," www.bls.gov/emp/ep_table_303.htm, accessed June 22, 2011.

5. Bureau of Labor Statistics, "Labor Force Statistics from the Current Population Survey," www.bls.gov/cps, accessed April 22, 2011.

6. Bureau of Labor Statistics, *Report on the American Workforce*, 69, www.bls.gov/opub/rtaw/pdf/chapter2.pdf.

7. Bureau of Labor Statistics, "Labor Force Statistics from the Current Population Survey, Unemployment Rate," April 10, 2011, http://data.bls.gov/pdq/SurveyOutputServlet?data_tool=latest_numbers&series_id=LNS14000000.

8. Bureau of Labor Statistics, "Local Area Unemployment Statistics," www.bls.gov/lau, accessed April 10, 2011.

9. Ibid.

10. See, for example, U.S. Chamber of Commerce, "The Impact of State Employment Policies on Job Growth: A 50 State Review," www.uschamber.com/reports/impact-state-employment-policies-job-growth-50-state-review, and Josh Biverns, "Why Do Some States Have Higher Unemployment Rates?" Economic Policy Institute, March 9, 2011, www.epi.org/publications/entry/why_do_some_states_have_higher_unemployment_rates.

11. "Nevada Leads the Nation in Unemployment," *Las Vegas Review-Journal*, June 19, 2010, www.lvrj.com/business/nevada-_x92-s-unemployment-rate-tops-in-nation-96647594.html.

12. Randy Ilg, "Long-Term Unemployment Experience of the Jobless," Bureau of Labor Statistics, www.bls.gov/opub/ils/summary_10_05/long_term_unemployment.htm.

13. Catherine Rampell, "Average Length of Unemployment Reaches High," *Economix* blog, *New York Times*, March 4, 2011, http://economix.blogs.nytimes.com/2011/03/04/average-length-of-unemployment-reaches-high-of-37-1-weeks.

14. See data at http://pewsocialtrends.org/pubs/759/how-the-great-recession-has-changed-life-in-america.

15. International Monetary Fund and International Labour Organization, "The Challenges of Growth, Employment, and Social Cohesion," 2010, 25, www.osloconference2010.org/discussionpaper.pdf.

16. Don Peck, "How a New Jobless Era Will Transform America," *Atlantic*, March 2010, www.theatlantic.com/doc/print/201003/jobless-america-future?x=47&y=1.

17. Quoted in ibid.

18. Bureau of Labor Statistics News Release, "Occupational and Wages by Ownership—May 2009," July 27, 2010, www.bls.gov/news.release/ocwag2.htm.

19. Ibid.

20. Ibid.

21. Ibid.

22. Ibid. This BLS data does not include people who are self-employed.

23. Bureau of Labor Statistics News Release, "Occupational Employment and Wages News Release," May 17, 2010, www.bls.gov/news.release/ocwage.htm.

24. Ibid.

25. Ibid.

26. Bureau of Labor Statistics News Release, "Usual Weekly Earnings of Wage and Salary Workers, Second Quarter 2010," Table 1, July 20, 2010, www.bls.gov/news.release/pdf/wkyeng.pdf.

27. Ibid.

28. See, for example, David Autor, "The Polarization of Job Opportunities in the U.S. Job Market," Center for American Progress and the Hamilton Project, August 2010, www.americanprogress.org/issues/2010/04/pdf/job_polarization.pdf.

29. Michael Mandel, "A Lost Decade for Jobs," *Bloomberg Businessweek*, June 23, 2009, www.businessweek.com/the_thread/economicsunbound/archives/2009/06/a_lost_decade_f.html.

30. Central Intelligence Agency, *World Factbook*, "Country Comparison—Unemployment Rate," https://www.cia.gov/library/publications/the-world-factbook/rankorder/2129rank.html.

31. OECD, "Implementing the OECD Jobs Strategy: Lessons from Member Countries' Experience," July 18, 2002, www.oecd.org/dataoecd/42/52/1941687.pdf.

32. See, for example, Rachel Donadio, "Europe's Young Grow Agitated over Future Prospects," *New York Times*, January 1, 2011, www.nytimes.com/2011/01/02/world/europe/02youth.html?_r=1& pagewanted=all.

33. Narayana Kocherlakota, "Inside the FOMC," speech, Federal Reserve Bank of Minneapolis, August 17, 2010, www.minneapolisfed.org/news_events/pres/speech_display.cfm?id=4525.

34. Economic Policy Institute, "Debunking the Claim of Structural Unemployment," September 10, 2010, www.epi.org/analysis_and_opinion/entry/debunking_the_theory_of_structural_unemployment.

35. Payscale.com, "The PayScale Index: Trends in Compensation," January 4, 2011, www.payscale.com/payscale-index.

CHAPTER 4: CHURN, BABY, CHURN

1. www.elvis-presley-biography.com/ReleaseMe.htm.

2. www.elvis.com/about-the-king/biography_/1935_1954/1935-1957_page_2.aspx.

3. www.elvis.com/news/detail.aspx?id=2436.

4. Mark Zandi, "Using Unemployment Insurance to Help Americans Get Back to Work," testimony before the Senate Finance Committee, April 14, 2010, http://finance.senate.gov/imo/media/doc041410mztest.pdf.

5. Mark Thoma, "What Level of Job Creation will Sustain Full Employment?" *Economist's View*, November 3, 2006, http://economistsview.typepad.com/economistsview/2006/11/what_level_of_j.html.

6. Phil Izzo, "Scarred Job Market Expected to Weigh on Economy," *Wall Street Journal*, October 8, 2009, http://online.wsj.com/article/SB125494927938671631.html.

7. Congressional Budget Office, "Budget and Economic Outlook Fiscal Years 2011–2021," January 2011, www.cbo.gov/doc.cfm?index=12039.

8. Susan K. Urahn, "Susan K. Urahn on Federal Stimulus for the States," Pew Center on the States, www.pewcenteronthestates.org/report_detail.aspx?id=49858.

9. Robert Longley, "Economic Stimulus Package—Tax Cuts: Tax Help for Families," Individuals and Small Businesses, About.com, May 2, 2011, http://usgovinfo.about.com/od/moneymatters/a/ecstimtaxes.htm.

10. See, for example, Robert Pollin et al.,"Green Recovery: A Program to Create Good Jobs and Start Building a Low-Carbon Economy," Political Economy Research Institute, University of Massachusetts, Amherst, September 2008, www.peri.umass.edu/fileadmin/pdf/other_publication_types/peri_report.pdf.

11. Rasmussen Reports, "National Survey of 1,000 Likely Voters," July 1, 2010, www.rasmussenreports.com/public_content/politics/toplines/pt_survey_toplines/july_2010/toplines_stimulus_i_july_1_2010.

12. CBO Director's Blog, "Estimated Impact of ARRA on Employment and Economic Output from July 2010 Through September 2010," November 24, 2010, http://cboblog.cbo.gov/?p=1617.

13. See, for example, Patrice Hill, "Unemployment Rate Stays at 9.6 Per-

cent: Economy Lost Overall 95,000 Jobs in September Due to Government Layoffs," *Washington Times*, October 8, 2010, www.washingtontimes.com/news/2010/oct/8/unemployment-stayed-96-percent-last-month.

14. See, for example, David Kestenbaum,"Economists Question Keynes-Inspired Stimulus," NPR, August 6. 2010, www.npr.org/templates/story/story.php?storyId=129031780.

15. Congressional Budget Office, *Report on the Troubled Asset Relief Program*, March 2010, Page 5, www.cbo.gov/ftpdocs/112xx/doc11227/03-17-TARP.pdf.

16. www.gm.com/corporate/about/company.jsp.

17. Neil King Jr. and Sharon Terlep, "GM Collapses into Government's Arms," *Wall Street Journal*, June 2, 2009, http://online.wsj.com/article/SB124385428627671889.html.

18. See, for example, John Kenneth Galbraith, "When Nissan Had a Better Idea," *New York Times*, October 26, 1986, www.nytimes.com/books/98/03/15/home/halberstam-reckoning.html.

19. Martin Felstein, "A Chapter for Detroit to Open," *The Auto Industry's Future*, American Enterprise Institute, November 24, 2008, www.aei.org/issue/28979.

20. Office of the Special Inspector General for the Troubled Asset Relief Program, "Factors Affecting the Decisions of General Motors and Chrysler to Reduce Their Dealership Networks," July 19, 2010, www.sigtarp.gov/reports/audit/2010/Factors%20Affecting%20the%20Decisions%20of%20General%20Motors%20and%20Chrysler%20to%20Reduce%20Their%20Dealership%20Networks%207_19_2010.pdf.

21. Ibid, 29.

22. Kate Andersen Brower and Nicholas Johnston, "Obama Says Auto Industry Revival Justifies Bailouts," *Bloomberg Businessweek*, July 30, 2010, www.businessweek.com/news/2010-07-30/obama-says-auto-industry-revival-justifies-bailouts.html.

23. Nick Bunkley, "Resurgent G.M. Posts 2010 Profit of $4.7 Billion," *New York Times*, February 24, 2011, www.nytimes.com/2011/02/25/business/25auto.html.

24. W. Michael Cox and Richard Alm, "Creative Destruction," *The Concise Encyclopedia of Economics,* www.econlib.org/library/Enc/Creative Destruction.html.

25. Ibid.

26. *San Francisco Business Times*, "Cisco Replaces GM in Dow Jones," June 1, 2009, http://sanfrancisco.bizjournals.com/sanfrancisco/stories/2009/06/01/daily10.html.

27. Bureau of Labor Statistics, "New Quarterly Data from BLS on Business Employment Dynamics by Size of Firm Summary," Table B: "Number of Firms and Employment by Size Class, First Quarter 2005, Not Seasonally Adjusted," December 8, 2005, www.bls.gov/news.release/cewfs.nr0.htm.

28. Starbucks, "Our Heritage," www.starbucks.com/about-us/our-heritage.

29. Hoovers, "Starbucks Corporation," www.hoovers.com/company/Star bucks_Corporation/rhkchi-1-1njg4g.html.

30. "Alex Manoogian, 95; Perfected Design of Single-Handled Faucet," *New York Times*, July 13, 1996, www.nytimes.com/1996/07/13/us/ alexmanoogian-95-perfected-design-of-single-handled-faucet.html.

31. Masco, "Historical Timeline," www.masco.com/corporate_informa tion/historical_timeline/index.html.

32. Masco, "Corporate Information," www.masco.com/corporate_infor mation/index.html.

33. Funding Universe, "Apple Computer, Inc.," www.fundinguniverse .com/company-histories/Apple-Computer-Inc-Company-History.html.

34. Securities and Exchange Commission, Form 10-K, Apple, Inc. at http://phx.corporate-ir.net/External.File?item=UGFyZW50SUQ9Njc1 MzN8Q2hpbGRJRD0tMXxUeXBlPTM=&t=1.

35. Bureau of Labor Statistics, "Career Information: Child Care Worker," www.bls.gov/k12/help02.htm.

36. Bureau of Labor Statistics, "Teachers—Special Education," *Occupational Outlook Handbook, 2010–11 Edition*, www.bls.gov/oco/ocos070. htm#emply.

37. Ibid.

38. Department of Labor, Office of Disability Employment Policy, "Business Ownership—Cornerstone of the American Dream," www.dol .gov/odep/pubs/business/business.htm.

39. Dane Stangler and Robert E. Litan, "Where Will the Jobs Come From?" Ewing Marion Kauffman Foundation, November 2009, 6, www.kauffman.org/uploadedFiles/where_will_the_jobs_come_from .pdf.

40. Quoted in Robert J. Samuelson, "In the Aftermath of the Great Recession," *Washington Post*, January 4, 2010, www.washingtonpost.com/ wp-dyn/content/article/2010/01/03/AR2010010301810.html.

41. www.accountingcoach.com/online-accounting-course/20Xpg04 .html.

42. Kaiser Family Foundation, "New Health Reform Law: Patient Protection and Affordable Care Act," www.kff.org/healthreform/ upload/8061.pdf.

43. Joe Hadzima, MIT Enterprise Forum, http://web.mit.edu/e-club/ hadzima/how-much-does-an-employee-cost.html.

44. Paul M. Krawzak and Melissa S. Bristow, "Even Now, Firms Struggle to Find Skilled Staff," Kiplinger.com, December 21, 2009, www .kiplinger.com/businessresource/forecast/archive/even-now-firms-struggle-to-find-skilled-staff.html.

45. Bureau of Labor Statistics, "The 30 Fastest-Growing Occupations, 2008–18," Table 7, www.bls.gov/news.release/ecopro.t07.htm.

46. See, for example, Mike Elgan, "Why We Need More H-1B Workers," Datamation.com, March 11, 2009, http://itmanagement.earthweb .com/columns/executive_tech/article.php/3809876/Why-We-Need-More-H-1B-Workers.htm.

47. The Iowa Policy Project, referenced in Peter Coy, Michelle Conlin, and Moira Herbst, "Future of America's Workforce: Permatemps," *Bloomberg Businessweek*, January 10, 2010, www.msnbc.msn.com/id/34769831.

48. See, for example, recent national surveys, available at www.polling report.com/consumer2.htm.

49. Department of Commerce, Bureau of Economic Analysis, "Measuring the Economy: A Primer on GDP and the National Income and Product Accounts," September 2007, 1, www.bea.gov/national/pdf/nipa_primer.pdf.

50. Ibid.

51. Ibid.

52. National Bureau of Economic Research, "Business Cycle Dating Committee," September 20, 2010, www.nber.org/cycles/sept2010.html.

53. Congressional Budget Office, CBO's Economic Projections for Calendar Years 2010 to 2021, Table 2-A, January 2011, http://www.cbo.gov/ftpdocs/120xx/doc12039/EconomicTables%5B1%5D.pdf.

54. Congressional Research Service, "Unemployment and Economic Recovery," August 20, 2010, http://opencrs.com/document/R40925/2010-08-20.

55. Sewell Chan, "Economists See Signs of Stronger Recovery," *New York Times*, December 23, 2010, www.nytimes.com/2010/12/24/business/economy/24forecast.html?_r=1&hp.

56. Catherine Rampell, "Companies Spend on Equipment, Not Workers," *New York Times*, June 9, 2011, www.nytimes.com/2011/06/10/business/10capital.html?_r=1&hp=&pagewanted=print.

57. Catherine Rampell, "The Growing Underclass: Jobs Gone Forever," *New York Times*, January 28, 2010, http://economix.blogs.nytimes.com/2010/01/28/the-growing-underclass-jobs-gone-forever.

58. Congressional Budget Office, "Policies for Increasing Economic Growth and Employment in 2010 and 2011," January 2010, 9, http://cbo.gov/ftpdocs/108xx/doc10803/01-14-Employment.pdf.

59. Rampell, "The Growing Underclass."

60. Alexander J. Field, "Productivity," *The Concise Encyclopedia of Economics*, www.econlib.org/library/Enc/Productivity.html.

61. Ibid.

62. Paul M. Romer, "Compound Rates of Growth," *The Concise Encyclopedia of Economics*, www.econlib.org/library/Enc/EconomicGrowth.html.

63. Mary Amiti and Kevin Stiroh, "Is the United States Losing Its Productivity Advantage?" Federal Reserve Bank of New York, Current Issues in Economics and Finance, September 2007, www.ny.frb.org/research/current_issues/ci13-8/ci13-8.html.

64. Federal Reserve Bank of San Francisco, "The Productivity and Jobs Connection: The Long and the Short Run of It," *FRBSF Economic Letter*, July 16, 2004, www.frbsf.org/publications/economics/letter/2004/el2004-18.html.

65. Ibid.; Field, "Productivity."
66. U.S. Census Bureau, "Economic Indicators," www.census.gov/cgi-bin/briefroom/BriefRm.
67. Kimberly Amadeo, "Stock Market History: The Dow Jones Industrial Average Historical Facts," About.com, http://useconomy.about.com/od/stockmarketcomponents/a/Dow_History.htm.
68. Ibid.
69. Darrell Rigby, "Look Before You Lay Off," *Harvard Business Review*, April 2002, http://hbr.org/2002/04/look-before-you-lay-off/ar/1.
70. See "Lay Off the Layoffs: Our Overreliance on Downsizing Is Killing Workers, the Economy—and Even the Bottom Line," *Newsweek*, February 05, 2010, www.newsweek.com/2010/02/04/lay-off-the-layoffs .html, and Rigby, "Look Before You Lay Off."
71. Walmart, "Sam Walton, Our Founder," http://walmartstores.com/AboutUs/9502.aspx.
72. Walmart Corporate Fact Sheet, http://walmartstores.com/download/2230.pdf.
73. Ibid.
74. See, for example, Steven Greenhouse, "Wal-Mart and Critics Step Up Battle over Wages," *New York Times*, May 5, 2005, www.nytimes .com/2005/05/04/business/worldbusiness/04iht-walmart.html, and Walmart Watch, http://walmartwatch.org.
75. Walmart Corporate Fact Sheet, http://walmartstores.com/download/2230.pdf.
76. "Community," MarthaStewart.com, www.marthastewart.com/portal/site/mslo/menuitem.a869edc68b016ad593598e10d373a0a0/?vgnextoid=d77e95ea62d7f010VgnVCM1000005b09a00aRCRD&autonomy_kw=best%20from%20martha&rsc=header_6.
77. Corporate Information, "Martha Stewart Living Omnimedia, Inc.," www.corporateinformation.com/Company-Snapshot.aspx?cusip=573083102.
78. Frito-Lay North America, "Our History," www.fritolay.com/about-us/history.html.
79. Pepsico, "PepsiCo Delivers Strong Financial and Operating Results for Fourth Quarter and Full Year 2010," www.pepsico.com/Press Release/PepsiCo-Delivers-Strong-Financial-and-Operating-Results-for-Fourth-Quarter-and-F02102011.html.
80. Amazon.com, "Media Kits: Overview," http://phx.corporate-ir.net/phoenix.zhtml?c=176060&p=irol-mediaKit, and Academy of Achievement, "Jeffrey P. Bezos," www.achievement.org/autodoc/page/bez0bio-1.
81. Amazon.com, "Media Kits: Overview."
82. Yahoo! Finance, "Amazon.com Inc.," http://finance.yahoo.com/q/ks?s=AMZN.
83. Amazon.com, "Media Kits: Overview."

CHAPTER 5: WOULD BALANCING THE BUDGET HELP CREATE JOBS?

1. www.amiright.com/quotes/doors.shtml.
2. Patrick McGilligan, *Alfred Hitchcock: A Life in Darkness and Light* (Hoboken, NJ: Wiley, 2003), 19, http://books.google.com/books?id=B f5l0qtZabMC&pg=PA20&lpg=PA20&dq=Alfred+Hitchcock+being+pu nished+in+school&source=bl&ots=U8qlE5PL29&sig=X8WM04sIgN vpncZd5DM1uzM33c&hl=en#v=onepage&q&f=false.
3. *CBO Director's Blog*, "The Economic Outlook and Fiscal Policy Choices," September 28, 2010, http://cboblog.cbo.gov/?p=1427.
4. Office of Management and Budget, "Mid-Session Review: Budget of the United States Government, FY2012," Table 1: "Change in Deficits from the February Budget," July 23, 2010, www.whitehouse.gov/omb/ budget/MSR.
5. Congressional Budget Office, "The Long-Term Budget Outlook," June 10, 2010, www.cbo.gov/doc.cfm?index=11579.
6. Ibid.
7. Donna Smith, "Panel Says U.S. Can't Grow Its Way out of Deficits," Reuters.com, April 26, 2010, www.reuters.com/article/idUSTRE 63O1WP20100426.
8. Congressional Budget Office, "Current Budget Projections: An Update," August 2010, Table 1-2, www.cbo.gov/ftpdocs/117xx/doc11705 /BudgetProjections.pdf.
9. Scott Bittle and Jean Johnson, *Where Does the Money Go? Your Guided Tour to the Budget Crisis* (New York: HarperCollins, 2011), www.wheredoesthemoneygo.com.
10. Pew Research Center, "Public's Wish List for Congress—Jobs and Deficit Reduction," July 12, 2010, http://people-press.org/report/633.
11. Economix blog, "Moody's: Why the U.S. Is Still AAA," *New York Times*, August 8, 2011, http://economix.blogs.nytimes.com/2011/08/08/ moodys-why-the-u-s-is-still-aaa/.
12. Standard & Poor's, "United States of America Long-Term Rating Lowered To 'AA+' Due to Political Risks, Rising Debt Burden; Outlook Negative," August 5, 2011, www.standardandpoors.com/ratings/articles/ en/us/?assetID=1245316529563.
13. Congressional Budget Office, "Federal Debt and the Risk of a Fiscal Crisis," July 27, 2010, www.cbo.gov/doc.cfm?index=11659.
14. *CBO Director's Blog*, "Estimated Impact of ARRA on Employment and Economic Output from July 2010 through September 2010," November 24, 2010, http://cboblog.cbo.gov/?p=1617.
15. Alan Greenspan, "U.S. Debt and the Greece Analogy," *Wall Street Journal*, June 18, 2010, http://online.wsj.com/article/SB1000142405274870 4198004575310962247772540.html, and Kenneth Rogoff, "No Need for a Panicked Fiscal Surge," FT.com, July 20, 2010, www.ft.com/cms/ s/0/6571e6c8-93f5-11df-83ad-00144feab49a.html#axzz16amVvvh8.
16. Jennifer Pietras, "Austerity Measures in the EU—A Country-by-Country Table," European Institute, www.europeaninstitute.org/Special-G-

20-Issue-on-Financial-Reform/austerity-measures-in-the-eu.html, accessed June 10, 2011, and "VAT Increase Q&A," *Telegraph*, www.telegraph.co.uk/finance/newsbysector/retailandconsumer/8238514/VAT-increase-QandA-what-is-affected-and-how-does-the-tax-work.html, accessed June 24, 2011.

17. See, for example, Vicki Needham and Ian Swanson, "Momentum Builds for Extending All of President Bush's Tax Cuts," *The Hill*, September 9, 2010, http://thehill.com/blogs/on-the-money/domestic-taxes/117859-nebraskas-nelson-support-extending-all-tax-cuts.

18. See, for example, "Global Economic Policy: Austerity Alarm," *Economist*, July 1, 2010, www.economist.com/node/16485318.

19. Daniel J. Mitchell, "The Fallacy That Government Creates Jobs," Cato Institute, December 5, 2008, www.cato.org/pub_display.php?pub_id=9825.

20. Club for Growth, "Economic Philosophies: Pro-Growth Tax Policy," www.clubforgrowth.org/philosophy, accessed November 28, 2010.

21. See, for example, Mark Zandi, "A Federal Shutdown Could Derail the Economy," Moody's Analytics, February 28, 2011, www.economy.com/dismal/article_free.asp?cid=197630&src=wp.

22. Joe Klein, "Obsessed with the Deficit—and Ignoring the Economic Mess," *Time*, November 18, 2010, www.time.com/time/nation/article/0,8599,2031958,00.html.

23. Katrina vanden Heuvel, "Obama Needs a Budget to Match His Progressive Ideals," *Washington Post*, April 19, 2011, www.washingtonpost.com/opinions/the-budget-plan-progressives-should-embrace/2011/04/18/AF7gSG5D_story.html?hpid=z4.

24. Our Fiscal Security, "Investing in America's Economy: A Budget Blueprint for Economic Recovery and Fiscal Responsibility," November 29, 2010, www.ourfiscalsecurity.org/storage/Blueprint_ExecSumm.pdf.

25. Robert Kuttner, "What Planet Are Deficit Hawks Living On?" *Huffington Post*, November 14, 2010, www.huffingtonpost.com/robert-kuttner/what-planet-are-deficit-h_b_783308.html.

26. Zandi, "A Federal Shutdown Could Derail the Economy."

27. Paul Krugman, "March of the Peacocks," *New York Times*, January 29, 2010, www.nytimes.com/2010/01/29/opinion/29krugman.html.

28. Ibid.

29. Alice Rivlin, "Commentary: Suppose We Just Keep Borrowing," *Nightly Business Report*, PBS, November 15, 2010, www.brookings.edu/economics/rivlin_nbr.aspx.

30. Ibid.

31. Scott Bittle and Jean Johnson, *Where Does the Money Go? Your Guided Tour to the Budget Crisis* (New York: HarperCollins, 2011), 329.

CHAPTER 6: WOULD CUTTING TAXES
HELP CREATE JOBS?

1. Tax Policy Center, *The Tax Policy Briefing Book*, "The Numbers: How Do U.S. Taxes Compare Internationally?" www.taxpolicycenter.org/briefing-book/background/numbers/international.cfm, accessed September 17, 2010.

2. World Economic Forum, *The Global Economic Competitiveness Report, 2010–2011*, www2.weforum.org/en/initiatives/gcp/Global%20Competitiveness%20Report/index.htm.

3. Ibid.

4. See, for example, David Leonhardt, "The Paradox of Corporate Taxes," *New York Times*, February 1, 2011, www.nytimes.com/2011/02/02/business/economy/02leonhardt.html?_r=1.

5. Tax Policy Center, "Tax Topics: Who Doesn't Pay Federal Taxes?" www.taxpolicycenter.org/taxtopics/federal-taxes-households.cfm, accessed September 10, 2010.

6. Casey B. Mulligan, "Why the Big Deal About Consumer Spending?" *Economix* blog, *New York Times*, March 9, 2011, http://economix.blogs.nytimes.com/2011/03/09/why-the-big-deal-about-consumer-spending/#more-103244.

7. Steven Pearlstein, "Put the Millionaires' Tax Money to Good Use," *Washington Post*, September 2, 2010, www.washingtonpost.com/wpdyn/content/article/2010/09/02/AR2010090205017.html.

8. Rick Newman, "Why Rich Consumers Matter More," *US News & World Report*, December 4, 2009, http://finance.yahoo.com/news/Why-Rich-Consumers-Matter-usnews-2047155287.html?x=0.

9. See, for example, Alan D. Viard, "The High-Income Rate Reductions: The Neglected Stepchild of the Bush Tax Cuts," American Enterprise Institute for Public Policy Research, September 2010, www.aei.org/outlook/100989.

10. www.rock-songs.com/songfacts/maggie-may.html; Stephen J. Dubner, "When the Rolling Stones Hit the Laffer Curve," Freakonomics, www.freakonomics.com/2010/11/01/when-the-rolling-stones-hit-the-laffer-curve.

11. Bruce Bartlett, "GOP 'No New Taxes' Position Is Rapidly Crumbling," *Fiscal Times*, May 13, 2011, www.thefiscaltimes.com/Columns/2011/05/13/GOP-Position-No-New-Taxes-Is-Rapidly-Crumbling.aspx?p=1.

12. See, for example, Steven Pearlstein, "Enough with the Economic Recovery: It's Time to Pay Up," *Washington Post*, June 11, 2010, www.washingtonpost.com/wp-dyn/content/article/2010/06/10/AR2010061004971.html?wpisrc=nl_pmheadline.

13. Federal Reserve Bank of San Francisco, "U.S. Household Deleveraging and Future Consumption Growth," *FRBSF Economic Letter*, May 15, 2009, www.frbsf.org/publications/economics/letter/2009/el2009-16.html.

14. See, for example, Mark Trumbull, "Can Americans Save Money—and

the Economy?" *Christian Science Monitor,* January 12, 2010, www .csmonitor.com/USA/2010/0112/Can-Americans-save-money-and-the-economy.

15. Joe Hadzima, MIT Enterprise Forum, http://web.mit.edu/e-club/ hadzima/how-much-does-an-employee-cost.html.

16. Congressional Budget Office, "Policies for Increasing Growth and Employment in 2010 and 2011," January 2010, 18, www.cbo.gov/ ftpdocs/108xx/doc10803/01-14-Employment.pdf.

17. Ibid.

18. Nouriel Roubini, "What America Needs Is a Payroll Tax Cut," *Washington Post,* September 17, 2010, www.washingtonpost.com/wp-dyn/ content/article/2010/09/16/AR2010091605846.html?wpisrc=nl_wonk.

19. Ibid.

20. Congressional Budget Office, "Policies for Increasing Growth and Employment in 2010 and 2011," 18.

21. Jia Lynn Yang, "Companies Pile Up Cash, But Remain Hesitant to Add Jobs," *Washington Post,* July 15, 2010, www.washington-post.com/wp-dyn/content/article/2010/07/14/AR2010071405960 .html?sid=ST2010082003981.

22. Gary T. Burtless, "Tax Cuts for New Hires: Not Yet Ready for Prime Time," Tax Policy Center, November 05, 2009, www.taxpolicycenter .org/publications/url.cfm?ID=1001344.

23. Quoted in Neil King Jr. and Gary Fields, "White House, Business Leaders Split on How to Create Jobs," *Wall Street Journal,* November 30, 2009, http://online.wsj.com/article/SB125954924769768987.html.

24. Quoted in David Kocieniewski, "Tax Cuts May Prove Better for Politicians than for Economy," *New York Times,* September 10, 2010, www .nytimes.com/2010/09/11/business/economy/11tax.html?_r=1&hp.

25. Steven Pearlstein, "Put the Millionaires' Tax Money to Good Use," *Washington Post,* September 2, 2010, www.washingtonpost.com/wpdyn /content/article/2010/09/02/AR2010090205017.html.

26. Bruce Bartlett, "The Case Against a Payroll Tax Cut," *New York Times,* August 30, 2011, http://economix.blogs.nytimes.com/2011/08/30/the-case-against-a-payroll-tax-cut/.

27. Ibid.

28. Mark Thoma, "Why Employment Might Not Fully Recover Until 2013," Moneywatch.com, November 10, 2009, http://moneywatch. bnet.com/economic-news/blog/maximum-utility/the-long-road-to-recovery-2/120.

CHAPTER 7: WOULD CUTTING
BUREAUCRACY HELP CREATE JOBS?

1. www.brainyquote.com/quotes/authors/r/ronald_reagan_2.html #ixzz17isMmqZn.

2. Encyclopedia of Alabama, "Kudzu," www.encyclopediaofalabama .org/face/Article.jsp?id=h-2483.

3. Southeast Exotic Pest Plant Council, *Invasive Plant Manual,* www

.se-eppc.org/manual/kudzu.html, and Ontario Invasive Plant Council, "Kudzu Vine: One of Ontario's Most Unwanted Invasive Plant Species," www.ipaw.org/invaders/OIPC_Un-wanted_Kudzu.pdf.

4. Southeast Exotic Pest Plant Council, *Invasive Plant Manual*, and Ontario Invasive Plant Council, "Kudzu Vine."

5. Dictionary.com, "Bureaucracy," http://dictionary.reference.com/browse/ bureaucracy, accessed December 13, 2010.

6. www.readprint.com/chapter-2800/Little-Dorrit-Charles-Dickens.

7. "Code of Federal Regulations: Main Page," www.gpoaccess.gov/cfr/ index.html, and James Gattuso, "Red Tape Rising: Regulatory Trends in the Bush Years," March 25, 2008, www.heritage.org/research/ reports/2008/03/red-tape-rising-regulatory-trends-in-the-bush-years.

8. Nicole V. Crain and W. Mark Crain, "The Impact of Regulatory Costs on Small Firms," Small Business Administration Office of Advocacy, September 2010, 13, www.sba.gov/advo/research/rs371tot.pdf.

9. Congressional Research Service, "CRS Issue Statement on Regulations and Rulemaking," January 6, 2010, http://pennyhill.net/docu ments/regulations_and_rulemaking.pdf.

10. Veronique de Rugy and Melinda Warren, "Expansion of Regulatory Budgets and Staffing Continues to Rise," Regulatory Report No. 13, Mercatus Center, George Mason University, 2009. Quoted in Crain and Crain, "The Impact of Regulatory Costs on Small Firms."

11. See, for example, Carl Bialik, "Small Business Regulatory 'Burden' Is Tough to Quantify," *Wall Street Journal*, January 29, 2011, http:// online.wsj.com/article/SB10001424052748703956604576110083387842092.html.

12. Crain and Crain, "The Impact of Regulatory Costs on Small Firms," 6.

13. Ibid., 7.

14. Ibid.

15. Varshney and Associates, "Cost of State Regulations on California Small Businesses Study," September 2009, 31, www.sba.ca.gov/ Cost%20of%20Regulation%20Study%20%20Final.pdf.

16. "Corporate Filing Forms," Arizona Corporation Commission, www .azcc.gov/divisions/corporations/filings/forms/index.asp.

17. *Gale Encyclopedia of US History*, "Government Regulation of Business," www.answers.com/topic/government-regulation-of-business.

18. Ibid.

19. Time.com, "Top 10 Product Recalls," March 24, 2010, www.time.com/ time/specials/packages/completelist/0,29569,1908719,00.html.

20. See, for example, Ronald N. Johnson and Gary D. Libecap, *The Federal Civil Service and the Problem of Bureaucracy* (Chicago: University of Chicago Press, 1994), www.nber.org/chapters/c8632.pdf.

21. Dane Stangler and Robert E. Litan, "Where Will the Jobs Come From?" Ewing Marion Kauffman Foundation, November 2009, 6, www.kauffman.org/uploadedFiles/where_will_the_jobs_come_from .pdf.

22. Global Entrepreneurship Monitor, *2007 Executive Report*, www.gem

consortium.org/download/1313784340950/GEM_2007_Executive_
Report.pdf.

23. "Snipping Off the Shackles," *Economist*, November 4, 2010, www
.economist.com/node/17419783.

24. Adam Bluestein and Amy Barrett, "Cutting Incorporation Bureaucracy," *Inc.*, July 1, 2010, www.inc.com/magazine/20100701/cutting-
incorporation-bureaucracy.html.

25. Scott Shane, "Time to Cut Red Tape for Startups," *Bloomberg Businessweek*, December 29, 2009. www.businessweek.com/smallbiz/con
tent/dec2009/sb20091228_471395.htm.

26. Bluestein and Barrett, "Cutting Incorporation Bureaucracy."

27. James L. Gattuso and Stephen A. Keen, "Red Tape Rising: Regulation in the Obama Era," Heritage Foundation Backgrounder, April 8,
2010, http://origin.heritage.org/Research/Reports/2010/03/Red-Tape-
Rising-Regulation-in-the-Obama-Era.

28. Terry McDonald, "New EPA Regulation May Discourage U.S. Biomass
Projects," Renewable Energy World, December 10, 2010, www.renew
ableenergyworld.com/rea/news/article/2010/12/new-epa-regulation-
may-discourage-u-s-biomass-projects.

29. Crain and Crain, "The Impact of Regulatory Costs on Small Firms," 7.

30. Ibid., 17.

31. Ibid., 20.

32. Alana Semuels, "Employers Embracing Automation to Cut Costs, Red
Tape," *Los Angeles Times*, October 14, 2010, www.dallasnews.com/
sharedcontent/dws/bus/stories/101410dnbusautomation.1fd3d3d.html.

33. Ibid.

34. IRS, Publication 926, www.irs.gov/publications/p926/ar02.html#en_
US_publink100086758.

35. Ibid.

36. Government Accountability Office, "Chemical Regulation: Observa-
tions on Improving the Toxic Substances Control Act," December 2,
2009, www.gao.gov/new.items/d10292t.pdf.

37. OMB Watch, "EPA Suspends Chemical Reporting," May 17, 2011,
www.ombwatch.org/node/11668.

38. See, for example, Carl Bialik, "Small Business Regulatory 'Burden'
Is Tough to Quantify," *Wall Street Journal*, January 29, 2011, http://
online.wsj.com/article/SB100014240527487039566045761100833878 4
2092.html.

39. Ruth Ruttenberg and Associates for Public Citizen, *Not Too Costly
After All: An Examination of the Inflated Cost-Estimates of Health,
Safety, and Environmental Protections*, February 2004, 1, www.citizen
.org/documents/ACF187.pdf.

40. Ibid.

41. Ibid.

42. Catherine Rampell, "Lax Oversight Caused Crisis, Bernanke Says,"
New York Times, January 3, 2010, www.nytimes.com/2010/01/04/busi
ness/economy/04fed.html?hp.

43. The dissenting panel members included former congressman Bill Thomas of California; Keith Hennessey, who served on the National Economic Council; Douglas Holtz-Eakin, a former head of the Congressional Budget Office; and Peter Wallison, a fellow at the American Enterprise Institute. See Carrie Bay, "Crisis Panel's GOP Members Fault Government for Housing Bubble," *Dallas Morning News*, December 16, 2010, www.dsnews.com/articles/crisis-panels-gop-members-fault-government-for-housing-bubble-2010-12-16.

44. Bay, "Crisis Panel's GOP Members Fault Government."

45. See, for example, Neil King Jr., "Sunstein's Ideas at Work in U.S. Policy," *Wall Street Journal*, July 6, 2009, http://online.wsj.com/article/SB124683695891298003.html.

46. Testimony of J. Robert Shull, deputy director for auto safety and regulation policy, Public Citizen, before the House Committee on Government Reform re: "The Small Business Paperwork Amnesty Act," September 26, 2006, 9, www.citizen.org/documents/testimony-hr5242.pdf.

47. "Initiatives: Al Gore, Vice President of the United States," http://clinton3.nara.gov/WH/EOP/OVP/initiatives_bottom.html.

48. June Hollis, "Kudzu: Weed or Valuable Plant?" About.com, http://geography.about.com/library/misc/uckudzu.htm.

49. City and County of San Francisco, "Labor Standards Enforcement," www.sfgsa.org/index.aspx?page=431.

50. Kaiser Family Foundation, "Focus on Health Reform: Summary of New Health Reform Law," April 15, 2011, www.kff.org/healthreform/upload/8061.pdf.

CHAPTER 8: WOULD REVIVING MANUFACTURING HELP CREATE JOBS?

1. www.mrpopculture.com/files/html/june01-1964. Kodak, "History of KODAK Cameras," www.kodak.com/global/en/consumer/products/techInfo/aa13/aa13pg2.shtml.

2. Mike Dickinson, "Kodak Employment Here Dips Below 7,500," *Rochester Business Journal*, February 1, 2010, www.rbj.net/article.asp?aID=182834, and Barbara Hagenbaugh, "U.S. Manufacturing Jobs Fading Away Fast," *USA Today*, December 12, 2002, www.usatoday.com/money/economy/2002-12-12-manufacture_x.htm.

3. Dickinson, "Kodak Employment Here Dips Below 7,500."

4. http://en.wikipedia.org/wiki/Kodachrome_(song), and www.sing365.com/music/lyric.nsf/kodachrome-lyrics-paul-simon/929e50784b2fafbb4825698a000b5995.

5. www.kodak.com/eknec/PageQuerier.jhtml?pq-path=2709&pq-locale=en_US&gpcid=0900688a80b4e692.

6. Carol J. Loomis, Patricia Neering, and Christopher Tkaczyk, "The Sinking of Bethlehem Steel," CNN Money, April 5, 2004, http://money.cnn.com/magazines/fortune/fortune_archive/2004/04/05/366339/index.htm_and http://chestertontribune.com/Business/bethlehem_steel_will_never_see_1.htm.

7. "Last Major US Jeans Manufacturer to Close US Plant," Designer Denim, November 7, 2006, 6, http://designer.denim.in.th/article.php/LastMajorUSJeansManufacturertofactory.

8. Industrial College of the Armed Forces, *Final Report: Manufacturing Industry*, spring 2009, 16–17, www.ndu.edu/icaf/programs/academic/industry/reports/2009/pdf/icaf-is-report-manufacturing-2009.pdf.

9. Bureau of Labor Statistics, *International Comparisons of Annual Labor Force Statistics, Adjusted to U.S. Concepts, 10 Countries, 1970–2009*, Table 2-8, www.bls.gov/fls/flscomparelf/employment.htm#table2_4.

10. The Manufacturing Institute, *The Facts About Modern Manufacturing*, 2010, 2, http://institute.nam.org/view/2001005059420889929.

11. Ibid., 11.

12. Ibid., 15.

13. Catherine Rampell, "Manufacturing Around the World," *Economix* blog, *New York Times*, November 16, 2009, http://economix.blogs.nytimes.com/2009/11/16/manufacturing-around-the-world.

14. See, for example, David Brauer, *Factors Underlying the Decline in Manufacturing Employment Since 2000*, Congressional Budget Office, December 23, 2008, www.cbo.gov/doc.cfm?index=9749&type=1; Robert Reich, "Manufacturing Jobs Are Never Coming Back," Forbes.com, May 28, 2009, www.forbes.com/2009/05/28/robert-reich-manufacturing-business-economy.html; and James Sherk, "Technology Explains Drop in Manufacturing Jobs," Heritage Backgrounder No. 2476, October 12, 2010, www.heritage.org/research/reports/2010/10/technology-explains-drop-in-manufacturing-jobs.

15. Michaela D. Platzer and Glennon J. Harrison, "The U.S. Automotive Industry: National and State Trends in Manufacturing Employment," Congressional Research Service, August 3, 2009, 8, http://digitalcommons.ilr.cornell.edu/cgi/viewcontent.cgi?article=1671&context=key_workplace.

16. Barbara Hagenbaugh, "U.S. Manufacturing Jobs Fading Away Fast," *USA Today*, December 12, 2002, www.usatoday.com/money/economy/2002-12-12-manufacture_x.htm.

17. Platzer and Harrison, "The U.S. Automotive Industry," 8.

18. http://cameracharts.1001noisycameras.com.

19. David Brauer, *Factors Underlying the Decline in Manufacturing Employment Since 2000*, Congressional Budget Office, December 23, 2008, 1, www.cbo.gov/doc.cfm?index=9749&type=1.

20. John Tschetter, *Exports Support U.S. Jobs*, U.S. Department of Commerce, 2010, 6, http://trade.gov/publications/pdfs/exports-support-american-jobs.pdf.

21. Platzer and Harrison, "The U.S. Automotive Industry," 8.

22. See, for example, NBC News/*Wall Street Journal* poll, November 11–15, 2010, available at http://online.wsj.com/public/resources/documents/WSJpoll111710.pdf.

23. See, for example, Alan S. Blinder, "Free Trade," *The Concise Encyclopedia of Economics*, www.econlib.org/library/Enc/FreeTrade.html.

24. See, for example, Ambassador Terry Miller, "Productivity Growth, Not Trade, Is Cutting Manufacturing Jobs," Heritage Foundation Web Memo No. 1709, November 27, 2007, www.heritage.org/research/reports/2007/11/productivity-growth-not-trade-is-cutting-manufacturing-jobs, and Josh Bivens, "Shifting Blame for Manufacturing Job Loss: Effect of Rising Trade Deficit Shouldn't Be Ignored," EPI Briefing Paper No. 149, April 8, 2004, www.epi.org/publications/entry/briefingpapers_bp149.

25. Ben S. Bernanke, "Embracing the Challenge of Free Trade: Competing and Prospering in a Global Economy," speech for the Montana Economic Development Summit, May 1, 2007, www.federalreserve.gov/newsevents/speech/bernanke20070501a.htm.

26. Tschetter, *Exports Support U.S. Jobs*. U.S. Department of Commerce, 2010, 6, http://trade.gov/publications/pdfs/exports-support-american-jobs.pdf.

27. Office of the United States Trade Representative, *The 2010 National Trade Estimate Report: Key Elements*, www.ustr.gov/about-us/press-office/fact-sheets/2010/march/-2010-national-trade-estimate-report-key-elements.

28. Ibid.

29. Ibid.

30. Office of the United States Trade Representative, *Free Trade Agreements*, www.ustr.gov/trade-agreements/free-trade-agreements.

31. The Manufacturing Institute, *The Facts About Modern Manufacturing*, 2010, 40, http://institute.nam.org/view/2001005059420889929.

32. Dean Zerbe, "America's Destructive Tax Code," Forbes.com, May 27, 2009, www.forbes.com/2009/05/27/tax-policy-manufacturing-business-zerbe.html.

33. Richard Wolf, "Obama's Fiscal Commission Votes 11–7 for Deficit Cuts," *USA Today*, December 3, 2010, www.usatoday.com/communities/theoval/post/2010/12/obamas-fiscal-commission-wins-majority-for-deficit-cuts/1.

34. Manufacturing Institute, *Facts About Modern Manufacturing*, 41.

35. Ibid.

36. See, for example, David Leonhardt, "The Paradox of Corporate Taxes," *New York Times*, February 1, 2011, www.nytimes.com/2011/02/02/business/economy/02leonhardt.html, and Brady Dennis, "What's Next for Tax Code? Geithner Discusses Overhaul with Wide Range of Groups," *Washington Post*, February 3, 2011, www.washingtonpost.com/wp-dyn/content/article/2011/02/03/AR2011020306584.html.

37. Manufacturing Institute, *Facts About Modern Manufacturing*, 45.

38. Richard McCormack, "The United States Is Not the Place to Build a New Factory, According to Manufacturing Council Members," *Manufacturing & Technology News*, December 30, 2010, www.manufacturingnews.com/news/newss/manufacturingcouncil123.html.

39. Industrial College of the Armed Forces Report, *Final Report: Manufacturing Industry*, Spring 2010, 1, www.ndu.edu/icaf/programs/aca

demic/industry/reports/2010/pdf/icaf-is-report-manufacturing-2010
.pdf.

40. Ibid., 8.
41. Ibid., 10.
42. Ibid., 19.
43. James Sherk, "Technology Explains Drop in Manufacturing Jobs," Heritage Backgrounder No. 2476, October 12, 2010, 7, www.heri tage.org/research/reports/2010/10/technology-explains-drop-in-manufacturing-jobs.
44. Bureau of Labor Statistics, *International Comparisons of Annual Labor Force Statistics, Adjusted to U.S. Concepts, 10 Countries, 1970–2009,* Table 2-8, www.bls.gov/fls/flscomparelf/employment.htm#table2_8.
45. Economy in Crisis, "Economic Solutions," www.economyincrisis.org/ issue/solutions.
46. Louis Uchitelle, "Ron Bloom Is Obama's Manufacturing Emissary," *New York Times,* September 9, 2010, www.nytimes.com/2010/09/10/ business/economy/10manufacture.html.
47. *"Time* Essay: Peril: The New Protectionism," *Time,* December 6, 1971, www.time.com/time/magazine/article/0,9171,877501,00.html.
48. Sherk, "Technology Explains Drop in Manufacturing Jobs," 8.
49. Ibid., 1.
50. Robert B. Reich, "Manufacturing Jobs Are Never Coming Back," Forbes.com, May 28, 2009, www.forbes.com/2009/05/28/robert-reich-manufacturing-business-economy.html.
51. Ibid.
52. Sherk, "Technology Explains Drop in Manufacturing Jobs," 7.
53. HistoryCentral.com, "Remarks by President Clinton, President Bush, President Carter, President Ford, and Vice President Gore in Signing of NAFTA Side Agreements," September 14, 1993, www.historycen tral.com/Documents/Clinton/SigningNaFTA.html.
54. Ibid.
55. White House adviser Ron Bloom quoted in Louis Uchitelle, "Ron Bloom Is Obama's Manufacturing Emissary," *New York Times,* September 9, 2010, www.nytimes.com/2010/09/10/business/economy/10 manufacture.html.

CHAPTER 9: WOULD IMPROVING EDUCATION HELP CREATE JOBS?

1. *Rising Above the Gathering Storm, Revisited: Rapidly Approaching Category 5* (Washington, DC: National Academies Press, 2010), 71, www .nap.edu/catalog.php?record_id=12999#description.
2. Ibid., 18.
3. Dow Jones Industrial Average, "History of the Dow," www.djindexes .com/mdsidx/downloads/brochure_info/Dow_Jones_Industrial_Aver age_Brochure.pdf.
4. R. Hira, "U.S. Workers in a Global Job Market," *Issues in Science and*

Technology, spring 2009, www.issues.org/25.3/hira.html. Cited in *Rising Above the Gathering Storm, Revisited* (2010), 7.

5. *Rising Above the Gathering Storm: Energizing and Employing America for a Brighter Economic Future* (Washington, DC: National Academies Presss, 2007), 1–2, www.nap.edu/catalog/11463.html.

6. See, for example, Philip Mattera, *Your Tax Dollars at Work . . . Offshore,* Good Jobs First, 2004, www.washtech.org/reports/TaxDollars AtWork/offshoring_execsum_finalpdf.pdf.

7. www.whitehouse.gov/the-press-office/2010/08/09/remarks-president-higher-education-and-economy-university-texas-austin.

8. U.S. Department of Labor, Office of the Secretary, "The Workforce," www.dol.gov/oasam/programs/history/herman/reports/futurework/report/chapter1/main.htm, accessed October 30, 2010.

9. Jennifer C. Day and Andrea E. Curry, "Educational Attainment in the United States: March 1998 (Update)," P20-513, October 1998, U.S. Department of Commerce, Economics and Statistics Administration, Census Bureau, www.census.gov/population/socdemo/education/tablea-02.txU.S. Cited in U.S. Department of Labor, "The Workforce," www.dol.gov/oasam/programs/history/herman/reports/futurework/report/chapter1/main.htm, accessed October 30, 2010.

10. *Rising Above the Gathering Storm, Revisited* (2010), 11.

11. U.S. Department of Education, National Center for Education Statistics, "Fast Facts: How Does Achievement of American Students Compare to Students in Other Countries?" http://nces.ed.gov/fastfacts/display.asp?id=1.

12. U.S. Department of Labor, Office of the Secretary, "The Workforce," www.dol.gov/oasam/programs/history/herman/reports/futurework/report/chapter1/main.htm, accessed October 30, 2010.

13. Day and Curry, "Educational Attainment in the United States."

14. Jean Johnson and Jon Rochkind with Amber N. Ott and Samantha DuPont, *With Their Whole Lives Ahead of Them: Myths and Realities About Why So Many Students Fail to Finish College,* Public Agenda, 2009, 2, www.publicagenda.org/files/pdf/theirwholelivesaheadofthem.pdf.

15. *Rising Above the Gathering Storm, Revisited* (2010), 38.

16. National Conference of State Legislatures, "American Higher Education in Urgent Need of Reform, State Legislators Say," November 27, 2006, www.ncsl.org/default.aspx?tabid=12835.

17. *Rising Above the Gathering Storm, Revisited* (2010), 8.

18. Ibid., 7.

19. See, for example, the chart "Education Pays" from the Bureau of Labor Statistics, www.bls.gov/emp/ep_chart_001.htm, accessed April 24, 2011.

20. *Rising Above the Gathering Storm* (2007), 3.

21. David Leonhardt, "Chance and Circumstance," *New York Times,* November 28, 2008, www.nytimes.com/2008/11/30/books/review/Leonhardt-t.html.

22. Bill Gates, remarks before the Committee on Science and Technology,

United States House of Representatives, March 12, 2008, www.micro soft.com/presspass/exec/billg/speeches/2008/congress.mspx.

23. *Rising Above the Gathering Storm, Revisited* (2010), 20.

24. White House Fact Sheet, "Building American Skills by Strengthening Community Colleges," 2010, www.whitehouse.gov/sites/default/files/ White_House_Summit_on_Community_Colleges_Fact_Sheet.pdf.

25. College Board, "About Us," www.collegeboard.com/about/index.html.

26. College Board, *The Complete College Agenda Report 2010*, http://com pletionagenda.collegeboard.org/sites/default/files/reports_pdf/Prog ress_Report_2010.pdf.

27. "College Preparation Checklist," http://studentaid.ed.gov/students/ publications/checklist/main.html.

28. "College for All," Inside Higher Ed, February 25, 2010, www.inside highered.com/news/2009/02/25/obama.

29. Alliance for Excellent Education, "The High Cost of Low Educational Performance: New OECD Report Finds That 25 Point Increase in PISA Scores Could Lead to $40.6 Trillion Increase in United States's GDP," *Straight A's: Public Education Policy and Progress*, January 25, 2010, www.all4ed.org/events/WebinarHighCostLowEducationalPerfor mance011910.

30. College Board, *The Complete College Agenda Report 2010*.

31. American Enterprise Institute, Education Outlook Series, www.aei .org/outlooksBinder?page=1&bid=100015.

32. Vivek Wadhwa and Robert E. Litan, "Viewpoint: Turning Research into Inventions and Jobs," *Bloomberg Businessweek*, September 20, 2009, www .businessweek.com/technology/content/sep2009/tc20090918_628309 .htm.

33. Ibid.

34. Michael Rizzo, "Mission Not to Accomplish," Inside Higher Ed, July 28, 2009, www.insidehighered.com/views/2009/07/28/rizzo.

35. Ibid.

36. Paul Krugman, "Degrees and Dollars," *New York Times*, March 6, 2011, www.nytimes.com/2011/03/07/opinion/07krugman.html?_r=1.

37. Quoted in Noah Berger, "Are Too Many Students Going to College?" *Chronicle of Higher Education*, November 8, 2009, http://chronicle .com/article/Are-Too-Many-Students-Going-to/49039/?sid=cc&utm_ source=cc&utm_medium=en.

38. Bureau of Labor Statistics, *Occupational Outlook Handbook, 2010–11 Edition*, "Aircraft and Avionics Equipment Mechanics and Service Technicians," www.bls.gov/oco/ocos179.htm.

39. Bureau of Labor Statistics, *Occupational Outlook Handbook, 2010–11 Edition*, "Structural and Reinforcing Iron and Metal Workers," www .bls.gov/oco/ocos215.htm.

40. Government Accountability Office, "Multiple Employment and Training Programs," January 2011, www.gao.gov/new.items/d1192.pdf.

41. Manufacturing Institute News Alert, "Responding to President Obama's Call to Action on Skills, Manufacturing Institute and

NAM Announce Goal to Credential Half-a-Million Students for the Manufacturing Workforce," June 8, 2011, http://institute.nam.org/view/2001115925285732691.

42. www.brainyquote.com/quotes/quotes/k/kurtvonneg103944.html.

CHAPTER 10: WOULD A MAJOR NATIONAL INFRASTRUCTURE PROJECT HELP CREATE JOBS?

1. "325,000 See Mayor Dedicate Airport to World Service," *New York Times,* October 16, 1939, http://graphics8.nytimes.com/packages/pdf/topics/WPA/39_10_16.pdf.

2. Ibid.

3. "Times Topics: The Great Depression," *The New York Times,* accessed December 27, 2010, http://topics.nytimes.com/top/reference/times topics/subjects/g/great_depression_1930s/index.html.

4. Ibid.

5. Ibid.

6. Thomas J. DiLorenzo, "The Myth of Government Job Creation," Cato Policy Analysis No. 48, February 19, 1984, www.cato.org/pubs/pas/pa048.html.

7. Jodie T. Allen, "How a Different America Responded to the Great Depression," Pew Research Center, December 14, 2010, http://pewresearch.org/pubs/1810/public-opinion-great-depression-compared-with-now.

8. www.education.com/reference/article/great-depression-deal-19291939/?page=3.

9. Michael Grabell and Christopher Weaver, "The Stimulus Plan: A Detailed List of Spending," *ProPublica,* February 13, 2009, www.propublica.org/special/the-stimulus-plan-a-detailed-list-of-spending#stim_transportation.

10. American Society of Civil Engineers, *2009 Report Card for America's Infrastructure,* Executive Summary, 7, www.infrastructurereport card.org/sites/default/files/RC2009_exsummary.pdf.

11. Ibid., 2.

12. Ibid., 3.

13. Ibid., 1.

14. Scott Bittle and Jean Johnson, *Who Turned Out the Lights? Your Guided Tour to the Energy Crisis* (New York: HarperCollins, 2009), 188–92.

15. "Beyond Stimulus: Toward a New Transportation Agenda for America," Miller Center of Public Affairs, September 2009, http://miller center.org/policy/transportation.

16. Ibid.

17. "New Jersey Governor Says He Won't Budge on Scrapped Rail Tunnel," *New York Times,* October 27, 2010, www.nytimes.com/2010/10/28/nyregion/28tunnel.html.

18. Michael Lind, "Unemployment: Going Beyond Short-Term Fixes," Salon.com, November 30, 2009, www.salon.com/news/opinion/feature/2009/11/30/jobs.

19. James Heintz, Robert Pollin, and Heidi Garrett-Peltier, "How Infrastructure Investments Support the U.S. Economy," Political Economy Research Institute, University of Massachusetts, January 2009, 3, www.peri.umass.edu/236/hash/efc9f7456a/publication/333.
20. Ibid., 28.
21. Paul Krugman, "The Jobs Imperative," *New York Times*, November 30, 2009, www.nytimes.com/2009/11/30/opinion/30krugman.html?_r=2.
22. Ibid.
23. See, for example, Tad DeHaven, "White House's Job Creation Figures," Cato Institute, July 16, 2010, www.downsizinggovernment.org/whitehouses-job-creation-figures; Karen Campbell, "Did the Stimulus Create Jobs? White House Economic Report Is Unclear," Web Memo No. 2700, Heritage Foundation, February 8, 2010, www.heritage.org/Research/Reports/2010/02/Did-the-Stimulus-Create-Jobs-White-House-Economic-Report-Is-Unclear; and Recovery.gov, "Track the Money: Jobs Summary—National," www.recovery.gov/Transparency/RecipientReportedData/Pages/JobSummary.aspx.
24. See, for example, Harold Meyerson, "Stimulus Funds on the Slow Track," *Washington Post*, December 29, 2010, www.tampabay.com/opinion/columns/billions-remain-unspent-in-slow-motion-stimulus/1142477; and The Foundry blog, "Morning Bell: $787 Billion in Stimulus, Zero Jobs 'Created or Saved,'" January 12, 2010, http://blog.heritage.org/2010/01/12/morning-bell-787-billion-in-stimulus-zero-jobs-created-or-saved.
25. CBO Director's Blog, "Estimated Impact of ARRA on Employment and Economic Output from July 2010 through September 2010," November 24, 2010, http://cboblog.cbo.gov/p=1617.
26. DeHaven, "White House's Job Creation Figures."
27. Ben Lieberman, "Green Jobs: Environmental Red Tape Cancels Out Job Creation," Web Memo No. 2795, Heritage Foundation, February 4, 2010, www.heritage.org/Research/Reports/2010/02/Green-Jobs-Environmental-Red-Tape-Cancels-Out-Job-Creation.
28. Ibid.
29. Lilly Library Collections, Indiana University, "More About the WPA," accessed June 26, 2011, www.indiana.edu/~liblilly/wpa/wpa_info.html.

CHAPTER 11: WOULD CLOSING THE GAP BETWEEN RICH AND POOR HELP CREATE JOBS?

1. www.wisebread.com/64-funny-inspiring-and-stupid-money-quotes-from-famous-people.
2. Gordon Rayner, "We'll Manage Without Butlers or Servants, Say Prince William and Kate Middleton," *Telegraph*, December 27, 2010, www.telegraph.co.uk/news/newstopics/theroyalfamily/8207022/Well-manage-without-butlers-or-servants-say-Prince-William-and-Kate-Middleton.html.
3. Ibid.
4. Wes Allison, "Income Gap Is Big, But So Are Its Causes," PolitiFact

.com, December 14, 2007, http://politifact.com/truth-o-meter/article/2007/dec/14/PFA_incomegap.

5. Harris Interactive, "Global Poll Finds That Gap Between Rich and Poor Expected to Increase in Next Five Years," April/May 2008, www.harrisinteractive.com/NewsRoom/HarrisPolls/Financial Times/tabid/449/ctl/ReadCustom%20Default/mid/1512/ArticleId/195/Default.aspx.

6. U.S. Census Bureau, "Income: Income Inequality," Table F-3: "Mean Income Received by Each Fifth and Top 5 Percent of Families, All Races: 1966 to 2009," www.census.gov/hhes/www/income/data/historical/inequality/index.html.

7. Associated Press, "Census: Income Gap Between Rich and Poor Got Wider in 2009," *USA Today*, October 1, 2010, www.usatoday.com/money/economy/2010-09-28-census-income-gap_N.htm.

8. U.S. Census Bureau, "Income, Poverty and Health Insurance Coverage in the United States 2009: Summary of Findings," September 16, 2010, www.census.gov/newsroom/releases/archives/income_wealth/cb10-144.html.

9. U.S. Census Bureau, "Poverty: Highlights, 2009," www.census.gov/hhes/www/poverty/about/overview/index.html.

10. Sarah Anderson, Chuck Collins, Sam Pizzigoti, and Kevin Shih, "CEO Pay and the Great Recession," Institute for Policy Studies, September 1, 2010, 3, www.ips-dc.org/files/2433/EE-2010-web.pdf.

11. Ibid.

12. Scott A. Hodge, "Tax Burden of Top 1% Now Exceeds That of Bottom 95%," Tax Foundation, July 29, 2009, www.taxfoundation.org/blog/show/24944.html.

13. CIA, *World Factbook*, "Field Listing: Distribution of Family Income—Gini Index," https://www.cia.gov/library/publications/the-world-factbook/fields/2172.html.

14. Ibid.

15. See, for example, Allison, "Income Gap Is Big, but So Are Its Causes," and Hal R. Varian, "Economic Scene: Many Theories on Income Inequality, but One Answer Lies in Just a Few Places," *New York Times*, September 21, 2006, www.nytimes.com/2006/09/21/business/21scene.html?_r=1.

16. McKinsey Global Institute, *Changing the Fortunes of America's Workforce: A Human Capital Challenge*, June 2009, www.mckinsey.com/mgi/reports/freepass_pdfs/changing_fortunes/Changing_fortunes_of_Americas_workforce.pdf.

17. Ira T. Kay and Steven van Putten, "Executive Pay: Regulation vs. Market Competition," Cato Institute Policy Analysis No. 619, September 10, 2008, www.cato.org/pub_display.php?pub_id=9621.

18. Varian, "Economic Scene: Many Theories on Income Inequality."

19. McKinsey Global Institute, *Changing the Fortunes of America's Workforce*.

20. Timothy Noah, "The United States of Inequality: Did Immigration

Create the Great Divergence?" *Slate*, September 7, 2010, www.slate
.com/id/2266025/entry/2266506.

21. Michaela D. Platzer and Glennon J. Harrison, "The U.S. Automotive
Industry: National and State Trends in Manufacturing Employment,"
Congressional Research Service, August 3, 2009, 8, http://digitalcom
mons.ilr.cornell.edu/cgi/viewcontent.cgi?article=1671&context=key_
workplace.

22. Bureau of Labor Statistics News, "Union Members in 2010," January
21, 2011, www.bls.gov/news.release/pdf/union2.pdf.

23. David Card, "The Effect of Unions on Wage Inequality in the U.S.
Labor Market," *Industrial and Labor Relations Review*, January 2001,
http://emlab.berkeley.edu/~card/papers/union-wage-ineq.pdf.

24. See, for example, James Sherk, "Performance-Based Pay Driving
Increase in Inequality," Heritage Foundation Web Memo No. 1505, June
13, 2007, http://s3.amazonaws.com/thf_media/2007/pdf/wm1505.pdf.

25. David Bolchover, "Why High Pay Is Bad for Capitalism," *Huffington
Post*, March 28, 2010, www.huffingtonpost.com/david-bolchover/why-
high-pay-is-bad-for-c_b_516385.html.

26. Ibid.

27. Ibid.

28. Andrew Ross Sorkin, *Too Big to Fail* (New York: Penguin, 2011).

29. "Politics and Policy: Campaign '92," *Wall Street Journal*, January 15,
1992. Cited in Ira T. Kay and Steven van Putten, "Executive Pay: Regulation vs. Market Competition," Cato Institute Policy Analysis No. 619,
September 10, 2008, www.cato.org/pub_display.php?pub_id=9621.

30. Bob Herbert, "Winning the Class War," *New York Times*, November 26, 2010,
www.nytimes.com/2010/11/27/opinion/27herbert.html?_r=1&src=ISMR_
HP_LO_MST_FB.

31. Robert Reich, "The Root of Economic Fragility and Political Anger,"
Huffington Post, July 13, 2010, www.huffingtonpost.com/robert-reich/
the-root-of-economic-frag_b_644465.html?page=1.

32. Ibid.

33. See, for example, James Sherk, "Shared Prosperity: Debunking Pes-
simistic Claims About Wages, Profits, and Wealth," Heritage Foundation Backgrounder No. 1978, October 16, 2006, www.heritage
.org/Research/Reports/2006/10/Shared-Prosperity-Debunking-
Pessimistic-Claims-About-Wages-Profits-and-Wealth.

34. Ibid.

35. Ibid.; Will Wilkinson, "Thinking Clearly about Economic Inequality,"
Cato Institute Policy Analysis No. 640, July 14, 2009, www.cato.org/
pub_display.php?pub_id=10351.

36. Sherk, "Shared Prosperity."

37. See, for example, Skyla Freeman, "Americans Are Obsessed with
Stuff," Human Events, August 10, 2010, www.humanevents.com/arti
cle.php?id=38479.

38. Michael D. Tanner, "The Income Gap and Deficit Reduction: Will
Balancing the Federal Budget Mean Increasing Inequality Between

the Rich and the Poor?" *New York Times*, November 30, 2010, www
.nytimes.com/roomfordebate/2010/11/30/will-deficit-reduction-
increase-income-inequality-in-the-us/income-inequality-is-the-
wrong-focus-for-government-policy.

39. Ibid.

40. OECD, "Income Inequality and Poverty Rising in Most OECD Countries," October 21, 2008, www.oecd.org/document/25/0,3343,en _2649
_201185_41530009_1_1_1_1,00.html.

41. Sue Sturgis, "Real 'Norma Rae' Dies of Cancer After Insurer Delayed
Treatment," Institute for Southern Studies, September 14, 2009, www
.southernstudies.org/2009/09/real-norma-rae-dies-of-cancer-after-
insurer-delayed-treatment.html.

42. Jonathan Kim, "Re-Thinking Norma Rae: A Union Icon Falls Fight-
ing the Healthcare Industry," *Huffington Post*, September 16, 2009,
www.huffingtonpost.com/jonathan-kim/rethinking-emnorma-
raeem_b_287552.html.

43. Bureau of Labor Statistics, "Union Members—2010," January 21, 2011,
www.bls.gov/news.release/pdf/union2.pdf.

44. "The Changing Union Movement," excerpted from *Outline of the U.S.
Economy*, Bureau of International Information Programs, September
15, 2009, www.america.gov/st/business-english/2009/September/2009
0916115359ebyessedo0.2755091.html.

45. Ibid.

46. Bureau of Labor Statistics, "Union Members—2010."

47. See, for example, Newt Gingrich, speech before the Institute for
Policy Innovation's Reclaiming Liberty Event, November 11, 2010,
www.americansolutions.com/economy/2010/11/speaker-gingrichs-
remarks-at-the-institute-for-policy-innovation.php, and Brady Dennis and Peter Wallsten, "Obama Joins Wisconsin's Budget Battle,
Opposing Republican Anti-Union Bill," *Washington Post*, February 18,
2011, www.washingtonpost.com/wp-dyn/content/article/2011/02/17/
AR2011021705494.html.

48. James Sherk, "What Unions Do: How Labor Unions Affect Jobs and
the Economy," Heritage Foundation Backgrounder No. 2275, May
21, 2009, Page 1, www.heritage.org/Research/Reports/2009/05/What-
Unions-Do-How-Labor-Unions-Affect-Jobs-and-the-Economy.

49. Ibid.

50. Eric Pryne, "Boeing to Fight NLRB Complaint on 787 South Carolina
Plant," *Seattle Times*, April 21, 2011, http://seattletimes.nwsource.com/
html/businesstechnology/2014824566_charleston21.html.

51. Jim McNerney quoted in Pryne, "Boeing to Fight NLRB Complaint."

52. Jim Trauger quoted in Pryne, "Boeing to Fight NLRB Complaint."

53. Bureau of Labor Statistics, "Union Members—2010."

54. Dennis and Wallsten, "Obama Joins Wisconsin's Budget Battle."

55. "*Time* Poll Results: Americans' Views on Teacher Tenure, Merit Pay
and Other Education Reforms," *Time*, September 9, 2010, www.time
.com/time/nation/article/0,8599,2016994,00.html.

56. Public Agenda and Learning Point Associates, Supporting Teaching Talent, 2009, www.publicagenda.org/pages/supporting-teacher-talent-view-from-Generation-Y-topline#q40.
57. Ibid.
58. Paul Krugman, "The Mellon Doctrine," *New York Times*, March 31, 2011, www.nytimes.com/2011/04/01/opinion/01krugman.html?_r=1.
59. U.S. Chamber of Commerce, "U.S. Chamber Study Shows States Could Create Nearly 750,000 Jobs and 50,000 New Businesses by Streamlining Employment Regulations," March 2, 2011, www.uschamber.com/press/releases/2011/march/us-chamber-study-shows-states-could-create-nearly-750000-jobs-and-50000-ne.
60. Lawrence Mishel and Heidi Shierholz, "The Sad but True Story of Wages in America," Economic Policy Institute, March 15, 2011, www.epi.org/publications/entry/the_sad_but_true_story_of_wages_in_america.
61. Krugman, "The Mellon Doctrine."
62. U.S. Chamber of Commerce, "The Impact of State Employment Policies on Job Growth: A 50-State Review, 2011," www.workforcefreedom.com/sites/default/themes/wfi/images_wfi/CH-10-0247_Book-Combined_M6.pdf#page=1.
63. Bureau of Labor Statistics, "Local Area Unemployment Statistics," www.bls.gov/web/laus/laumstrk.htm.
64. Bureau of Labor Statistics, "Overview of BLS Wage Data by Area and Occupation," www.bls.gov/oes/current/oessrcst.htm.

CHAPTER 12: WILD CARD #1

1. N. Gregory Mankiw, "Economic View: Emerging Markets as Partners, Not Rivals," *New York Times*, February 13, 2011, www.nytimes.com/2011/02/13/business/13view.html.
2. Cato Institute, "Beyond Exports: A Better Case for Free Trade," *Free Trade Bulletin*, January 31, 2011, www.cato.org/pub_display.php?pub_id=12741.
3. *Washington Post* poll conducted January 13–17, 2011, www.washingtonpost.com/wp-dyn/content/article/2011/01/28/AR2011012801651.html.
4. NBC/*Wall Street Journal* poll, conducted Nov. 11–15, 2010, http://online.wsj.com/public/resources/documents/WSJpoll111710.pdf.
5. ABCNews.com, "Made in America: Meet the Usry Family," February 28, 2011, http://abcnews.go.com/US/MadeInAmerica/made-america-meet-usry-family/story?id=12950594.
6. Congressional Research Service, "Globalized Supply Chains and U.S. Policy," January 9, 2009, http://opencrs.com/document/R40167.
7. Congressional Budget Office, "Factors Underlying the Decline in Manufacturing Employment Since 2000," December 23, 2008, 2, Table 1, www.cbo.gov/doc.cfm?index=9749&type=1.
8. Mankiw, "Economic View: Emerging Markets as Partners, Not Rivals."

9. Congressional Research Service, "Globalized Supply Chains and U.S. Policy."

10. Ibid.

11. Congressional Research Service, "Offshoring and Job Insecurity Among U.S. Workers," August 6, 2007.

12. World Economic Forum, *The Global Competitiveness Report 2010–2011*, www3.weforum.org/docs/WEF_GlobalCompetitivenessReport_2010-11.pdf.

13. Andreja Pirc and Richard Vlosky, "A Brief Overview of the U.S. Furniture Industry," Louisiana Forest Products Development Center, July 14, 2010, www.lfpdc.lsu.edu/publications/working_papers/wp89.pdf.

14. Congressional Research Service, "Globalized Supply Chains and U.S. Policy."

15. Blinder, Alan, "Free Trade's Great, but Offshoring Rattles Me," May 6, 2007, www.washingtonpost.com/wp-dyn/content/article/2007/05/04/AR2007050402555_2.html.

16. Blinder, Alan, "How Many U.S. Jobs Might Be Offshorable?" CEPS Working Paper No. 142. March 2007, www.princeton.edu/ceps/workingpapers/142blinder.pdf.

17. Congressional Research Service, "Globalization, Worker Insecurity and Policy Approaches," January 20, 2010, www.fas.org/sgp/crs/misc/RL34091.pdf.

18. Economic Policy Institute, "A Policy Agenda for Offshoring," March 11, 2009, http://epi.3cdn.net/4732615b523cd3a3dd_o5m6bx6eq.pdf.

19. James K. Galbraith, "Why Populists Need to Re-think Trade," *American Prospect*, May 10, 2007, www.prospect.org/cs/articles?article=why_populists_need_to_rethink_trade.

20. Government Accountability Office, "Trade Adjustment Assistance: Changes Needed to Improve States' Ability to Provide Benefits and Services to Trade-Affected Workers," June 14, 2007, www.gao.gov/new.items/d07995t.pdf.

21. www.trade.gov/mas/ian/build/groups/public/@tg_ian/documents/webcontent/tg_ian_003364.pdf.

CHAPTER 13: WILD CARD #2

1. www.nytimes.com/2011/02/17/science/17jeopardy-watson.html.

2. www.npr.org/programs/morning/features/patc/johnhenry/index.html.

3. Slate.com, "The End: Why Projectionists Will Soon Be No More," December 6, 2010, www.slate.com/id/2266654.

4. www.theatlantic.com/business/archive/2011/04/americas-fastest-dying-business-its-mobile-homes/73336.

5. Pew Research Center for the People and the Press, "Cell Phones," accessed July 15, 2011, http://people-press.org/methodology/sampling/cell-phones.

6. See, for example, "Tough Times in the Porn Industry," *The Los Angeles Times*, August 10, 2009, www.latimes.com/news/local/la-fi-ct-porn10-2009aug10,0,3356050.story.

7. Bureau of Labor Statistics, *Occupational Outlook Handbook 2008–18*, accessed April 2, 2011, www.bls.gov/oco/oco2003.htm.

8. *30 Rock*, "Plan B," aired March 23, 2011, www.nbc.com/30-rock/video/plan-b/1315922.

9. Bureau of Labor Statistics, *Occupational Outlook Handbook 2008-18*, accessed April 2, 2011, www.bls.gov/oco/oco2003.htm.

10. John Markoff, "Armies of Expensive Lawyers, Replaced by Cheaper Software," *New York Times*, March 4, 2011, www.nytimes.com/2011/03/05/science/05legal.html.

11. "The Skill Content of Recent Technological Change: An Empirical Exploration," *Quarterly Journal of Economics*, November 2003, www.mitpressjournals.org/doi/abs/10.1162/003355303322552801.

12. Ibid.

13. Ibid.

14. "Why Jobs Haven't Come Back: The Technology Factor," *New York Times*, January 17, 2011, www.nytimes.com/roomfordebate/2011/01/17/why-jobs-havent-come-back/the-technology-factor.

15. Foreign Policy, "Ten Percent Unemployment Forever?" January 5, 2011, www.foreignpolicy.com/articles/2011/01/05/10_percent_unemployment_forever?page=0,0.

16. John Markoff, "Computer Wins on Jeopardy: Trivial It's Not," *New York Times*, February 16, 2011, www.nytimes.com/2011/02/17/science/17jeopardy-watson.html.

17. John Markoff, "A Fight to Win the Future: Computers vs. Humans," *New York Times*, February 14, 2011, www.nytimes.com/2011/02/15/science/15essay.html.

18. Martin Ford, "Anything You Can Do, Robots Can Do Better," *Atlantic*, February 14, 2011, www.theatlantic.com/business/archive/2011/02/anything-you-can-do-robots-can-do-better/71227.

19. John Markoff, "Armies of Expensive Lawyers, Replaced by Cheaper Software," *New York Times*, March 4, 2011, www.nytimes.com/2011/03/05/science/05legal.html.

20. Brad DeLong, "Grasping Reality with Both Hands," March 7, 2011, http://delong.typepad.com/sdj/2011/03/the-hollowing-out-of-the-us-income-distribution-under-the-pressure-of-technology.html.

21. Paul Krugman, "Degrees and Dollars," *New York Times*, March 4, 2011, www.nytimes.com/2011/03/07/opinion/07krugman.html.

22. Tyler Cowen, *The Great Stagnation: How America Ate All the Low-Hanging Fruit of Modern History, Got Sick, and Will (Eventually) Feel Better* (New York: Dutton, 2011), 85, www.amazon.com/Great-Stagnation-Low-Hanging-Eventually-ebook/dp/B004H0M8QS.

CHAPTER 14: WILD CARD #3

1. U.S. Department of Homeland Security, *Yearbook of Immigration Statistics 2010*, www.dhs.gov/files/statistics/publications/yearbook.shtm.

2. U.S. Census Bureau, *American Community Survey*, "Nativity Status and Citizenship in the United States, 2009," October 2010, www.cen

sus.gov/prod/2010pubs/acsbr09-16.pdf.

3. Congressional Budget Office, "The Role of Immigrants in the U.S. Labor Market: An Update," June 2010, www.cbo.gov/doc.cfm?index=11691.

4. Office of Immigration Statistics, U.S. Department of Homeland Security, "U.S. Legal Permanent Residents: 2010," March 2011, www.dhs .gov/xlibrary/assets/statistics/publications/lpr_fr_2010.pdf.

5. Ibid.

6. Federal Reserve Bank of San Francisco Economic Letter, "The Effect of Immigrants on U.S. Employment and Productivity," August 30, 2010, www.frbsf.org/publications/economics/letter/2010/el2010-26 .html.

7. Pew Hispanic Center, "Unauthorized Immigrant Population: National and State Trends, 2010," February 1, 2011, http://pewhispanic.org/ reports/report.php?ReportID=133.

8. Roger Lowenstein, "The Immigration Equation," *New York Times Magazine*, July 9, 2006, www.nytimes.com/2006/07/09/magazine/09IMM. html.

9. Congressional Research Service, "Immigration: The Effects on Low-Skilled and High-Skilled Native-Born Workers," April 13, 2010, http:// assets.opencrs.com/rpts/95-408_20100413.pdf.

10. Pew Hispanic Center, "Statistical Portrait of the Foreign-Born Population in the United States 2008," Table 23, http://pewhispanic.org/files/ factsheets/foreignborn2008/Table%2023.pdf.

11. National Research Council, "The New Americans: Economic, Demographic, and Fiscal Effects of Immigration," May 1997, www.nap.edu/ catalog.php?record_id=5779.

12. Congressional Research Service, "Immigration: The Effects on Low-Skilled and High-Skilled Native-Born Workers."

13. Federal Reserve Bank of San Francisco, "The Effect of Immigrants on U.S. Employment and Productivity."

14. Ibid.

15. See, for example Roy Mark, "Gates Rakes Congress on H1B Visa Cap," Datamation.com, April 27, 2005, http://itmanagement.earthweb.com/ career/article.php/3500986/Gates-Rakes-Congress-on-H1B-Visa-Cap .htm.

16. Congressional Research Service, "Foreign Science and Engineering Presence in U.S. Institutions and the Labor Force," October 28, 2010, http://fas.org/sgp/crs/misc/97-746.pdf.

17. Congressional Research Service, "Immigration: The Effects on Low-Skilled and High-Skilled Native-Born Workers."

18. Congressional Research Service, "Foreign Science and Engineering Presence in U.S. Institutions and the Labor Force."

19. National Research Council, "The New Americans: Economic, Demographic, and Fiscal Effects of Immigration," May 1997, www.nap.edu/ catalog.php?record_id=5779.

20. *EE Times*, "Immigrant Patent Filings Surge in U.S.," January 22, 2007, www.eetimes.com/electronics-news/4068554/Immigrant-patent-

filings-surge-in-U-S-.

21. Jennifer Hunt and Marjolaine Gauthier-Loiselle, "How Much Does Immigration Boost Innovation?" *American Economic Journal: Macroeconomics*, 2010, American Economic Association, www.nber.org/papers/w14312.pdf.

22. www.soc.duke.edu/globalengineering/pdfs/media/IntellectualProperty/americanspectator_needed.pdf.

23. Federal Reserve Bank of Richmond, "Research Spotlight: What Immigration Means for the Economy," second quarter 2010, www.richmondfed.org/publications/research/region_focus/2010/q2/pdf/research_spotlight.pdf.

24. Huma Khan, "'Shooting Itself in the Foot': Is U.S. Turning Away Entrepreneurs?" ABC News, April 21, 2010, http://abcnews.go.com/Politics/Business/economic-job-growth-driven-foreign-enterpreneurs-argue/story?id=10428413.

25. James Fallows, *Atlantic*, "How America Can Rise Again," January 2010, www.theatlantic.com/magazine/archive/2010/01/how-america-can-rise-again/7839.

26. "Hispanic Fast Facts: The Power of the Hispanic Market," http://ahaa.org/default.asp?contentID=161.

CHAPTER 15: WILD CARD #4

1. www.time.com/time/covers/0,16641,20110221,00.html?artId=20110221?contType=gallery?chn=covers.

2. U.S. Census Bureau, "Facts for Features: Oldest Boomers Turn 60," www.census.gov/newsroom/releases/archives/facts_for_features_special_editions/cb06-ffse01-2.html.

3. Civic Ventures, "Research in Brief—After the Recovery: Help Needed," 2010, www.encore.org/files/research/JobsBluestoneBriefLogos3-11-10.pdf.

4. Barry Bluestone and Mark Melnik, "After the Recovery: Help Needed," Civic Ventures, 2010, www.encore.org/files/research/JobsBluestoneBriefLogos3-11-10.pdf.

5. Alex Kowalski and Tom Keene, "Aging Baby Boomers Reduce U.S. Unemployment Rate, Matus Says," Bloomberg News, February 7, 2011, www.bloomberg.com/news/2011-02-07/aging-baby-boomers-reduce-u-s-unemployment-rate-matus-says-tom-keene.html.

6. Linda Levine, Congressional Research Service, *Retiring Boomers = A Labor Shortage?* January 30, 2008, summary, http://aging.senate.gov/crs/pension36.pdf.

7. Encore Careers, "The Coming Labor Shortage and How People in Encore Careers Can Help Solve It," www.encore.org/find/resources/after-recovery-help.

8. Bluestone and Melnik, "After the Recovery: Help Needed."

9. Encore Careers, "The Coming Labor Shortage."

10. Levine, *Retiring Boomers = A Labor Shortage?*

11. www.census.gov/newsroom/releases/archives/facts_for_features_spe cial_editions/cb06-ffse01-2.html.
12. www.census.gov/population/www/pop-profile/natproj.html.
13. Levine, *Retiring Baby Boomers = A Labor Shortage?*
14. Ibid., 2.
15. Ibid., 5.
16. Ibid., 10.
17. Ibid., 12.
18. American Association of Colleges of Nursing, "Nursing Shortage Fact Sheet," September 20, 2010, www.aacn.nche.edu/media/pdf/ NrsgShortageFS.pdf.
19. www.bls.gov/oco/cg/cgs035.htm.
20. American Association of Colleges of Nursing, "Nursing Shortage Fact Sheet."
21. www.bls.gov/oco/ocos083.htm#training.
22. American Association of Colleges of Nursing, "Nursing Shortage Fact Sheet."
23. Steven A. Sass, Courtney Monk, and Kelly Haverstick, "Workers' Response to the Market Crash—Save More, Work More?" Center for Retirement Research at Boston College, February 2010, 1, 3, http://crr .bc.edu/images/stories/Briefs/ib_10-3.pdf.
24. Ibid.
25. Employee Benefit Research Institute, Retirement Confidence Survey, 2020 RCS Fact Sheet #4: "Age Comparisons Among Workers," 3, www .ebri.org/pdf/surveys/rcs/2010/FS-04_RCS-10_Age.pdf.
26. Ibid.
27. Ibid.
28. Sass, Monk, and Haverstick, "Workers' Response to the Market Crash," 3.
29. EBRI, "2011 Retirement Confidence Survey—2011 Results," 31, www .ebri.org/pdf/surveys/rcs/2011/EBRI_03-2011_No355_RCS-11.pdf.
30. Barbara A. Butrica, Howard M. Iams, Karen E. Smith, and Eric J. Toder, "The Disappearing Defined Benefit Pension and Its Potential Impact on the Retirement Incomes of Baby Boomers," Social Security Administration, *Social Security Bulletin* 69, no. 3 (2009), www.ssa.gov/ policy/docs/ssb/v69n3/v69n3p1.html.
31. See, for example, Emily Brandon, "10 Ways Baby Boomers Will Reinvent Retirement," *US News*, February 16, 2010, http://money.usnews .com/money/retirement/articles/2010/02/16/10-ways-baby-boomers-will-reinvent-retirement, and Diana Jean Schemo, "Coming Boomer Pension Cuts: What Impact on the Economy?" Remapping Debate, October 12, 2010, www.remappingdebate.org/article/coming-boomer-pension-cuts-what-impact-economy?page=0,0.
32. Employee Benefit Research Institute, Retirement Confidence Survey, 2020 RCS Fact Sheet #4: "Age Comparisons Among Workers," 5, www .ebri.org/pdf/surveys/rcs/2010/FS-04_RCS-10_Age.pdf.
33. Pew Research Center, *Working After Retirement: The Gap Between*

Expectations and Reality, September 21, 2006, http://pewsocial trends.org/2006/09/21/working-after-retirement-the-gap-between-expectations-and-reality.

34. Steven Greenhouse, "65 and Up and Looking for Work," *New York Times*, October 24, 2009, www.nytimes.com/2009/10/24/business/economy/24older.html?_r=1&ref=economy.

35. Bureau of Labor Statistics, "Record Unemployment Among Older Workers Does Not Keep Them out of the Job Market," *Issues in Labor Statistics*, March 2010, www.bls.gov/opub/ils/pdf/opbils81.pdf.

36. Greenhouse, "65 and Up and Looking for Work."

37. Nelson D. Schwartz, "Easing Out the Gray-Haired—or Not," *New York Times*, May 27, 2011, www.nytimes.com/2011/05/28/business/economy/28worker.html.

38. "Schumpeter: The Silver Tsunami—Business Will Have to Learn How to Manage an Ageing Workforce," *Economist*, February 4, 2010, www.economist.com/node/15450864.

39. Ibid.

40. Jeffrey Joerres, "Aging Your Work Force: Keeping Older Employees Will Help Maintain Success," *Wall Street Journal*, April 9, 2009, http://online.wsj.com/article/SB123923040617102867.html.

41. www.conference-board.org/matureworker.

42. www.encore.org/about.

CHAPTER 16: FOURTEEN BIG IDEAS FOR CREATING MORE AND BETTER JOBS

1. Michael Mandel, "A Lost Decade for Jobs," *Bloomberg Businessweek*, June 23, 2009, www.businessweek.com/the_thread/economicsunbound /archives/2009/06/a_lost_decade_f.html.

2. See, for example, Steven Pearlstein, "Put the Millionaires' Tax Money to Good Use," *Washington Post*, September 2, 2010, www.washington post.com/wp-dyn/content/article/2010/09/02/AR2010090205017.html.

3. In 2011, the tax applied to worker salaries up to nearly $107,000. See the details at: www.ssa.gov/oact/progdata/taxRates.html.

4. Congressional Budget Office, "Policies for Increasing Growth and Employment in 2010 and 2011," January 2010, 18, www.cbo.gov/ftpdocs/108xx/doc10803/01-14-Employment.pdf.

5. Gary T. Burtless, "Tax Cuts for New Hires: Not Yet Ready for Prime Time," Tax Policy Center, November 5, 2009, www.taxpolicycenter.org/publications/url.cfm?ID=1001344.

6. Social Security and Medicare Boards of Trustees, "A Summary of the 2010 Annual Reports," www.ssa.gov/oact/TRSUM/index.html.

7. See, for example, "More Ideas for Creating Jobs," *Bloomberg Businessweek*, December 7, 2009, www.businessweek.com/magazine/content/09_49/b4158020735034.htm.

8. CNN/Opinion Research Corporation Poll, March 19–21, 2010, available at www.pollingreport.com/enviro.htm.

9. Nicole V. Crain and W. Mark Crain, "The Impact of Regulatory Costs

on Small Firms," Small Business Administration Office of Advocacy, September 2010, www.sba.gov/advo/research/rs371tot.pdf.

10. Rea Hederman and James Sherk, "Heritage Employment Report: 2009 Ends with More Job Losses," Heritage Web Memo No. 2748, January 8, 2010, www.heritage.org/research/reports/2010/01/heritage-employment-report-2009-ends-with-more-job-losses.

11. See, for example, William D. Ruckelshaus and and Christine Todd Whitman, "A Siege Against the EPA and Environmental Progress," *Washington Post*, March 24, 2011, www.washingtonpost.com/opinions/a-siege-against-the-epa-and-environmental-progress/2011/03/23/ABsuyeRB_story.html.

12. Energy Information Administration, Energy-in-Brief: "What Is the Electric Power Grid, and What Are Some Challenges It Faces?" October 20, 2009, www.eia.doe.gov/energy_in_brief/power_grid.cfm.

13. www.whitehouse.gov/the-press-office/president-obama-announces-34-billion-investment-spur-transition-smart-energy-grid.

14. Scott Bittle and Jean Johnson, *Who Turned Out the Lights? Your Guided Tour to the Energy Crisis* (New York: HarperCollins, 2009), 191.

15. See, for example, Andy Stone, "Smart Grid, Stupid Policy?" Forbes .com, January 29, 2009, www.forbes.com/2009/01/29/electricity-infrastructure-obama-business-energy-0129_smart_grid.html.

16. Bracken Hendricks, "Wired for Progress 2.0: Building a National Clean-Energy Smart Grid," Center for American Progress, April 1, 2009, www.americanprogress.org/issues/2009/04/wired_for_progress 2.0.html.

17. Bureau of Labor Statistics, "Wages: Minimum Wage," www.dol.gov/dol/topic/wages/minimumwage.htm.

18. Tiffany Williams, "Raise the Minimum Wage: Boosting It Would Help our Lowest-Paid Workers as Well as the Entire Economy," CommonDreams.org, November 22, 2010, www.commondreams.org/view/2010/11/22-9.

19. See, for example, Marilyn Geewax, "Does a Higher Minimum Wage Kill Jobs?" National Public Radio, April 23, 2011, www.npr .org/2011/04/24/135638370/does-a-higher-minimum-wage-kill-jobs.

20. Bureau of Labor Statistics, "Characteristics of Minimum Wage Workers: 2010," www.bls.gov/cps/minwage2010.htm.

21. Ibid.

22. Bureau of Labor Statistics, "Questions and Answers about the Minimum Wage," www.dol.gov/whd/minwage/q-a.htm.

23. www.commondreams.org/view/2010/11/22-9.

24. Michael J. Hicks, "Who Lost Jobs When the Minimum Wage Rose?" Bureau of Business Research, Ball State University, February 2010, Page 1, http://cms.bsu.edu/Academics/CentersandInstitutes/BBR/~/media/DepartmentalContent/MillerCollegeofBusiness/BBR/Publications/MinWage.ashx.

25. Pierre Cahuc and Andre Zylbergerg, *The Natural Survival of Work: Job Creation and Destruction in a Growing Economy* (Cambridge, MA: MIT Press, 2006).

26. Richard V. Burkhauser and Joseph J. Sabia, "Raising the Minimum Wage: Another Empty Promise to the Working Poor," EPI, August 2005, 1, http://epionline.org/studies/burkhauser_08-2005.pdf.

27. www.dol.gov/whd/minwage/america.htm.

28. www.bls.gov/web/laus/laumstrk.htm.

29. www.bls.gov/web/laus/laumstrk.htm and www.dol.gov/whd/minwage/america.htm.

30. Institute for Legal Reform, Issues Resource Center, www.institutefor legalreform.com/component/ilr_issues/29.html, accessed March 28, 2011.

31. Harris Interactive, "Small Businesses: The Threat of Lawsuits Impacts Their Operations: Summary of Findings," May 10, 2007, www.insti tuteforlegalreform.com/get_ilr_doc.php?id=1045.

32. Sandy Smith, "Insurance Industry Study: Asbestos Litigation Hurts Workers," December 5, 2002, http://ehstoday.com/news/ehs_ imp_36011.

33. Ibid.

34. See, for example, Helen Dewar, "Senate Republicans Seek to Limit Class-Action Suits," *Washington Post*, July 6, 2004, www.washington post.com/wp-dyn/articles/A29622-2004Jul5.html, and Rea Hederman and James Sherk, "Heritage Employment Report: 2009 Ends with More Job Losses," Heritage Web Memo No. 2748, January 8, 2010, www.heritage.org/research/reports/2010/01/heritage-employment-report-2009-ends-with-more-job-losses.

35. Adam Liptak, "Justices Rule for Wal-Mart in Class-Action Bias Case," *New York Times*, June 20, 2011, www.nytimes.com/2011/06/21/business/21bizcourt.html.

36. See, for example, Marie Gryphon, "Greater Justice, Lower Cost: How a 'Loser Pays' Rule Would Improve the American Legal System," Manhattan Institute for Policy Research, December 2008, www .manhattan-institute.org/html/cjr_11.htm.

37. Dan Slater, "The Debate over Who Pays Fees When Litigants Mount Attacks," WSJ.com, December 23, 2008, http://online.wsj.com/article/SB122999187816728533.html.

38. U.S. Chamber of Commerce, "Nine Out of Ten Voters Say 'Meritless' Lawsuits are a Serious Problem, Support Continued Legal Reforms," November 4, 2010, www.uschamber.com/press/releases/2010/novem ber/nine-out-ten-voters-say-%E2%80%98meritless-lawsuits-are-serious-problem-support.

39. See, for example, Marc I. Gross, "Loser-Pays—or Whose 'Fault' Is It Anyway," *Law and Contemporary Problems*, spring/summer 2001, www.law.duke.edu/shell/cite.pl?64+Law+&+Contemp.+Probs.+163+ (SpringSummer+2001), and Barbara Hart, "Lawsuits: Dismiss Class Actions—Con: Don't Underestimate an Opportunity for Justice," *Bloomberg Businessweek*, March 7, 2008, www.businessweek.com/debateroom/archives/2008/03/lawsuits_dismiss_class_actions.html.

40. Deborah L. Rhode, "Frivolous Litigation and Civil Justice Reform,"

Duke Law Journal, 2004, 456, https://www.law.duke.edu/journals/dlj/downloads/dlj54p447.pdf.

41. Ross Eisenbrey, "The Frivolous Case for Tort Law Change: Opponents of the Legal System Exaggerate Its Costs, Ignore Its Benefits," EPI Briefing Paper #157, May 16, 2005, www.epi.org/publications/entry/bp157.

42. www.bls.gov/k12/law02.htm#job.

43. www.bls.gov/k12/law05.htm.

44. Mike Lillis, "Economists Push for Federal Job-Sharing Program," *Washington Independent*, February 24, 2010, http://washingtonindependent.com/77609/economists-push-for-federal-job-sharing-program and Paul Krugman at www.nytimes.com/2009/11/13/opinion/13krugman.html.

45. Dean Baker, quoted in Lillis, "Economists Push for Federal Job-Sharing Program."

46. Dr. David G. Javitch, "The Pros and Cons of Job Sharing," *Entrepreneur*, November 10, 2006, www.entrepreneur.com/humanresources/employeemanagementcolumnistdavidjavitch/article170244.html.

47. Professor Don Boudreaux of George Mason University quoted in Tibor R. Machan, "Self-Correction in Markets v. Journalism," *A Passion for Liberty*, November 15, 2009, http://tibormachan.rationalreview.com/tag/don-boudreaux.

48. Lawrence Summers quoted in Catherine Rampell, "Should the Government Pay People to Work Less?" *Economix* blog, *New York Times*, November 10, 2009, http://economix.blogs.nytimes.com/2009/11/10/should-the-government-pay-people-to-work-less/#more-39851.

49. David Leonhardt, "The Paradox of Corporate Taxes," *New York Times*, February 1, 2011, www.nytimes.com/2011/02/02/business/economy/02leonhardt.html?_r=1.

50. Ibid.

51. "Simple, Fair, and Pro-Growth: Report of the President's Advisory Panel on Federal Tax Reform," November 2005, 34, www.taxpolicycenter.org/taxtopics/upload/Tax-Panel-2.pdf.

52. Howard Gleckman, "Should We Cut Corporate Taxes by Raising Rates on Investors?" Tax Policy Center, March 29, 2011, http://taxvox.taxpolicycenter.org/2011/03/29/should-we-cut-corporate-taxes-by-raising-rates-on-investors.

53. Tax Policy Center, "Historical Marginal Effective Tax Rates on Capital Income," January 13, 2004, www.taxpolicycenter.org/taxfacts/displayafact.cfm?Docid=323.

54. Congressional Budget Office, "Options for Responding to Short-Term Economic Weakness," January 2008, 13, www.cbo.gov/ftpdocs/89xx/doc8916/01-15-Econ_Stimulus.pdf.

55. U.S. Small Business Administration, "President Obama Issues Executive Order on Improving Regulation and Regulatory Review Advocacy Strongly Supports Focus on Small Business," January 18, 2011, www.sba.gov/advocacy/808/13743.

56. Conn Carroll, "Rising Tide of Government Regulation Faces Rising Opposition from American People," Heritage Foundation, October 29, 2010, http://blog.heritage.org/2010/10/29/rising-tide-of-government-regulation-faces-rising-opposition-from-american-people.

57. Ibid.

58. Nicole V. Crain and W. Mark Crain, "The Impact of Regulatory Costs on Small Firms," Small Business Administration Office of Advocacy, September 2010, 7, www.sba.gov/advo/research/rs371tot.pdf.

59. See, for example, Carl Bialik, "Small Business Regulatory 'Burden' Is Tough to Quantify," Wall Street Journal, January 29, 2011, http://online.wsj.com/article/SB10001424052748703956604576110083387842092.html.

60. Ruth Ruttenberg and Associates for Public Citizen, Not Too Costly After All: An Examination of the Inflated Cost-Estimates of Health, Safety, and Environmental Protections, February 2004, www.citizen.org/documents/ACF187.pdf.

61. Bureau of Labor Statistics, "Teachers—Special Education," Occupational Outlook Handbook, 2010–11 Edition, www.bls.gov/oco/ocos070.htm#emply.

62. Catherine Rampell, "A Decline in American Entrepreneurship," Economix blog, New York Times, March 31, 2011, http://economix.blogs.nytimes.com/2011/03/31/a-decline-in-american-entrepreneurship.

63. John Haltiwanger of the University of Maryland, cited in Adam Bluestein and Amy Barrett, "Revitalizing the American Dream," Inc., July 6, 2010, www.inc.com/magazine/20100701/revitalizing-the-american-dream.html.

64. CNN/Opinion Research Corporation poll, May 21–23, 2010, at http://politicalticker.blogs.cnn.com/2010/05/26/cnn-poll-support-for-border-crackdown-grows/.

65. See, for example, Quinnipiac University poll, August 31–September 7, 2010, http://articles.latimes.com/2010/sep/13/news/la-pn-immigration-mosque-poll-20100914.

66. Thomas J. Tierney, "American Higher Education: How Does It Measure Up for the 21st Century?" National Center for Public Policy and Higher Education, 2006, www.highereducation.org/reports/hunt_tierney/Hunt_Tierney.pdf.

67. Public Agenda, "Slip-Sliding Away: An Anxious Public Talks about Today's Economy," February 3, 2011, www.publicagenda.org/economy-and-american-dream-2011.

68. See, for example, Jamie P. Merisotis, "Higher Education Productivity in the 'New Era of Responsibility,'" speech before the Hartford Consortium for Higher Education, Lumina Foundation for Education, January 29, 2009, www.luminafoundation.org/about_us/president/speeches/2009-01-29.html, and Kresge Foundation, "Higher Education Productivity," www.kresge.org/index.php/what/education/productivity.

69. Jamie P. Merisotis and Stan Jones, "Degrees of Speed," Washington

Monthly, May/June 2010, www.washingtonmonthly.com/features/2010/1005.merisotis-jones.html.

70. The Ewing Marion Kauffman Foundation, *Starting Up Growth,* 2011, www.kauffman.org/uploadedFiles/tb2011_starting-up-growth.pdf.

71. Paul Krugman, "Degrees and Dollars," *New York Times,* March 6, 2011, www.nytimes.com/2011/03/07/opinion/07krugman.html.

72. *World Youth Report, 2003,* 58, www.un.org/esa/socdev/unyin/documents/ch02.pdf.

73. www.nationmaster.com/graph/edu_sch_lif_exp_tot-education-school-life-expectancy-total.

74. Jacob S. Hacker and Paul Pierson, "The Wisconsin Union Fight Isn't About Benefits. It's About Labor's Influence," *Washington Post,* March 6, 2011, www.washingtonpost.com/wp-dyn/content/article/2011/03/04/AR2011030402416.html.

75. See, for example, Marlon Bishop, "Union Fights 'Priscilla' Producers over Live Music," WNYC, May 16, 2011, http://culture.wnyc.org/articles/features/2011/may/16/musicians-union-fights-priscilla-producers.

76. Quoted in Aaron Bernstein, "Why America Needs Unions but Not the Kind It Has Now," *Bloomberg Businessweek,* May 23, 1994, www.businessweek.com/archives/1994/b337360.arc.htm.

77. Paul Krugman, "Free to Lose," *New York Times,* November 13, 2009, www.nytimes.com/2009/11/13/opinion/13krugman.html.

78. See, for example, James Sherk, "What Unions Do: How Labor Unions Affect Jobs and the Economy," Heritage Foundation Backgrounder No. 2275, May 21, 2009, www.heritage.org/Research/Reports/2009/05/What-Unions-Do-How-Labor-Unions-Affect-Jobs-and-the-Economy.

79. See, for example, Steven Malanga, "President Obama Won't Revive Unions," Manhattan Institute for Policy Research, August 13, 2008, www.manhattan-institute.org/html/miarticle.htm?id=2970.

80. Pew Research Center poll, February 22–March 1, 2011, http://people press.org/2011/03/03/section-4-opinions-of-labor-unions/.

81. Robert Wood Johnson Foundation, "Number of Americans with Employer-Sponsored Health Insurance Drops Significantly over Last Decade," June 21, 2011, www.rwjf.org/coverage/product.jsp?id=72527.

82. Geoff Lewis, "The Twilight of Employer-Sponsored Health Insurance?" *Small Business Review,* accessed June 25, 2011, http://smallbusiness review.com/for_the_boss/Employer_Sponsored_Health_Insurance.

83. Kaiser Family Foundation, "National Health Insurance: A Brief History of Reform Efforts in the U.S.," March 2009, www.kff.org/healthreform/upload/7871.pdf.

84. ABC News/*Washington Post* poll, February 4–8, 2010, available at www.pollingreport.com/health3.htm.

85. See, for example, Uwe E. Reinhardt. "Is Employer-Based Health Insurance Worth Saving?" *Economix* blog, *New York Times,* May 22, 2009, http://economix.blogs.nytimes.com/2009/05/22/is-employer-based-health-insurance-worth-saving.

86. Congressional Research Service, "Globalization, Worker Insecurity, and Policy Approaches," January 20, 2010, http://opencrs.com/document/RL34091/2010-01-20.

87. John Podesta and Sabina Dewan, "Just Jobs," Center for American Progress, October 7, 2010, www.americanprogress.org/issues/2010/10/just_jobs.html.

88. Cato Institute, "Fair Trade Coffee Enthusiasts Should Confront Reality," 2007, www.cato.org/pubs/journal/cj27n1/cj27n1-9.pdf.

89. Congressional Research Service, "Globalization, Worker Insecurity, and Policy Approaches," January 20, 2010, http://opencrs.com/document/RL34091/2010-01-20.

90. Kimberly Amadeo, "History of NAFTA," About.com, accessed June 25, 2011, http://useconomy.about.com/od/tradepolicy/p/NAFTA_History.htm.

91. www.visioncritical.com/wp-content/uploads/ . . . /2010.08.25_NAFTA.pdf.

CHAPTER 17: NOW WHAT? SEVEN GROUND RULES FOR MOVING FORWARD

1. www.brainyquote.com/quotes/authors/f/franklin_d_roosevelt_3.html#ixzz1KFd1Eiv3.

2. www.stlyrics.com/lyrics/grease/grease.htm.

3. Gallup poll, "Work and Workplace," accessed April 29, 2011, www.gallup.com/poll/1720/Work-Work-Place.aspx.

4. Public Agenda, "Slip-Sliding Away: An Anxious Public Talks About Today's Economy," February 3, 2011, www.publicagenda.org/economy-and-american-dream-2011.

5. Dennis Jacobe, "Americans Perceive Quality Jobs as Harder to Get in June," June 20, 2011, www.gallup.com/poll/148121/americans-perceive-quality-jobs-harder-june.aspx.

6. Gallup poll, "Economy," accessed April 29, 2011, www.gallup.com/poll/1609/Economy.aspx.